TOUCH THE
SEA

TOUCH THE SEA

DEE SCARR

To Peter + Ellen,
After great dives,
knowing you'll have many
more touching underwater
interactions!

Dee Scarr
+
the kissing grunts
+
our rescued spongecrab

Bonaire 4/94

TTS Publishing, 3717 Lankenau Rd.,
Philadelphia, PA 19131

Printed in the United States of America

ISBN # : 1-878663-05-4
Library of Congress # : 90-060547

To my parents, without whose support this book—and the experiences within it—would never have been possible.

Table of Contents

TOUCH THE
S E A

Introduction

Inner Space . . . Planet Ocean . . . Whatever it is called, the sea is our last frontier on this planet. Until recently, only the very edges of the sea were accessible for human exploration, and our knowledge of the deep oceans came from those glimpses we could catch at the seashore, and from what fishermen hauled up by luck or skill, with nets or dredges, in traps or on lines.

Now, thanks to the development of scuba (scuba is the acronym for Self-Contained Underwater Breathing Appara-

A queen angelfish stands out against a background of multicolored sponges.

A frogfish faces the camera.

tus), the fringe of the sea that we are able to explore has widened—and deepened—considerably. Scientists and capitalists, looking for new discoveries and new ways to make money, have turned their attention to the ocean's depths. They leave to sport scuba divers the intimacy and fascination of the coral reefs.

Exploring wilderness areas on land requires a great deal of patience. Most wild creatures are so shy that a naturalist must remain still for long periods of time before any animals appear at all; even more time is required before the animals take up their everyday activities.

The creatures of the sea—as long as they've not been hunted by divers—are much bolder than land creatures. They may disappear momentarily as the diver enters the water, but in a few moments, life on the reef—or the sand flat, or the kelp bed, or the mangrove swamp—settles back to what it was before the interruption. Brightly colored blue tangs nibble on algae. Pugnacious damselfish guard their territories from all intruders. Scorpionfish and frogfish and sea-horses stay in position, comfortable in their camouflage. Shrimps dance around flowerlike anemones while crabs remain out of sight, hidden by sea urchins or within the coral or by personal camouflages. The coral reef is a three-dimensional world and, unlike land wilderness areas, each dimension of the reef is readily available to the scuba-diving observer.

I have always been interested in the sea. As a child, before I had even heard of scuba diving, I spent hours around the canals in Miami, watching little snappers and grunts nibble pieces of hot dog off the tiniest hooks I could find. If I accidentally caught one, I released it.

Sometimes my Sundays would be brightened by a trip to the beach, and sometimes there would be sargassum sea-weed floating in the water. I found that on the seaweed itself lived fascinating inhabitants. While most kids were tossing the weed at each other, I was examining the clumps, lifting

Once they are coaxed out of hiding, octopi are rewarding to watch. Photo by Geri Murphy.

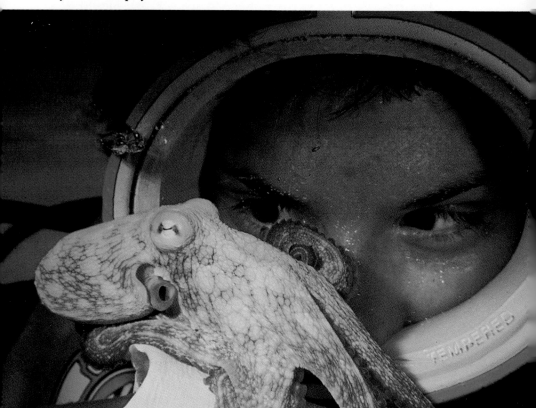

them gently from the water to see what was left behind. I found common little sargassum shrimp, golden-brown in color to match the weed, and sargassum crabs, marked with a pattern exactly like the spots and shadows on the weed itself. My favorite resident of the weed was the sargassumfish *(Histrio histrio)*, a small (up to four inches long) anglerfish perfectly camouflaged to match the weed. The anglerfish would sit upon the sargassum and wait for an unsuspecting grazer to try to nibble one of its imitation sargassum fins. Slurp! Gone was the grazer.

There were sargassum filefish, sargassum slugs, and even an occasional seahorse.

It was frustrating to find all these wonderful animals and then have no choice but release them back to the sea, so when I caught a particularly large sargassumfish one day, I summoned up all my courage and called the Miami Seaquarium. Would the oceanarium like a sargassumfish for its collection? It would!

How proud my friend and I felt as we ignored the ticket booths and headed, bucket in hands, for the office entrance. We were welcomed and invited to the back rooms—the inner sanctum not open to the public—to see our fish placed in a holding tank. Then we were set free to enjoy the Seaquarium grounds. It's funny how the sharks are more fierce, the dolphins more skillful, the ice cream more delicious, to a child whose entrance to that magical world was earned by adding a fish to the collection!

I talked to marine biologists at the Seaquarium and at the University of Miami Marine Lab, and wrote to others around the United States. I want to be a marine biologist, I told them. I want to study the sargassum community.

What were their responses? The logistics are too difficult, you'll need to have boats and go all the way into the Gulf Stream to get unpolluted weed. Why bother, they asked, with sargassum weed anyway? In essence, I was told that the topic wasn't worth any trouble, much less the trouble it

The delicate seahorse is a weak swimmer and is most often found anchored by its tail to a stalk of seaweed, a soft coral, or a sponge.

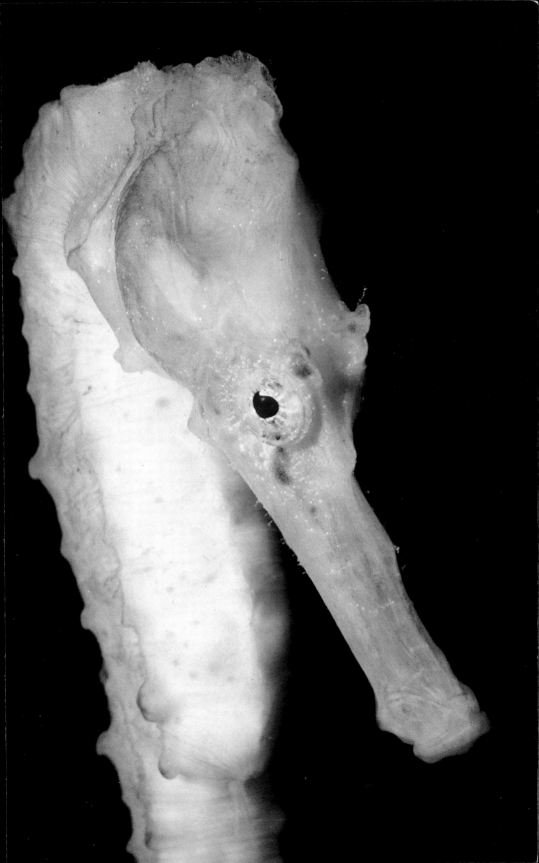

would require. I should spend my energies on matters of consequence.

So, like the flyer in *The Little Prince*, I gave up my idea. I went off to college, fell into the study of English literature, public speaking, and debate, earned bachelor's and master's degrees from the University of Florida, and returned to Miami to teach high school.

But still I visited Marineland of Florida, watched Jacques Cousteau's documentary specials on television, eagerly read about underwater exploration in *National Geographic,* enjoyed *Flipper,* and became scuba-certified the summer after my first year of teaching school.

Winter and summer school holidays I spent diving in the Florida Keys. In 1977, I took a leave of absence from teaching and in 1978, I was hired as a divemaster on San Salvador, in the Bahamas. I'd glide along underwater and think, "I can't believe I'm being *paid* to dive!" Almost two years later, I left San Salvador for Bonaire; one and a half years after that, I began my own "operation": *Touch the Sea,* a program of personalized dive guiding.

I once had a friend who burned out as a divemaster. He just didn't enjoy diving anymore; even bikini-watching wasn't enough to keep him from leaving the islands for a land-based, non diving-related job. I contrasted his way of approaching the reef to my own, and saw the beginning of the *Touch the Sea* concept.

The more I dive, the more fascination the sea holds for me. There are levels upon levels. . . . Levels of sight: a blenny living in a hole drilled by a worm through an encrusting sponge and around a scallop living on a pier piling. Levels of behavior: a "mob" of surgeonfish streaming from algae-clump to algae-clump, with attendant trumpetfish and jacks hunting prey distracted by the surgeonfish activities.

Observation alone is not enough for me, though. As you'll understand when you read this book, I want to belong to the undersea world, or at least to interact with it. What will result from interactions between humans and the creatures of the reef? If the result is greater understanding and caring, it will certainly be a positive one.

I hope you enjoy this book, and that in some way, whether by diving or even simply by virtue of an awakened interest, you, too, will want to touch the sea.

—Dee Scarr

I

~~~~~~

Tidepooling

I didn't expect much on my first tidepooling expedition on Bonaire—actually, I hadn't started out to go tidepooling at all. Two friends and I had planned a beach dive. One friend had forgotten something and gone back for it, the tide was low, and for lack of anything better to do I began to wade around. At first I saw mostly algae and the ever-present small tooth shells and periwinkles. Sand-colored blennies scooted and slithered from pool to pool. The more I looked, the more I saw. Rock-boring urchins, smaller than Ping-Pong balls, were so well cuddled into their niches that I didn't see them at all until I looked closely.

What was that sharp feeling on the bottom of my foot? Carefully, I inspected my foot and found a four-inch scorpionfish, spines erect. By carelessly walking along I had placed my foot right on top of the little guy, who—luckily for both of us—was protected by his crevice. He darted to a rock in the next tidepool, and I moved over and squatted down to have a better look at him. He let me stroke him gently.

My curiosity was piqued by the rock he had sought for protection. I turned it over and found that the space between rock and tidepool bottom was filled with dozens of tiny hermit crabs, all jumbled together. I replaced the rock carefully, and turned over some others. Under one was a chain moray eel in lovely colors of green and yellow. Under another, I found a tiny sea cucumber, so small it could have been called a sea pickle.

Part of the fascination of the sea, for me, is that each of its

A heron wades in the tidepools along the coast of Bonaire.

A nudibranch crawls over a pink sponge. Most nudibranchs are less than one inch long.

habitats, from tidepools to open blue water, has its own residents. Exploring tidepools is the simplest way to touch the sea. Tidepooling is so interesting, in fact, that many oceanariums now have "tidepool" displays, where visitors are encouraged to handle starfish, urchins, mollusks, and other tidepool inhabitants.

For tidepooling, you need: 1. a seashore upon which, at low tide, water is trapped in pools; 2. low tide; 3. rubber-soled enclosed shoes, such as dive booties or canvas sneakers; 4. a friend to share your discoveries with; 5. gloves, if you are expecting to turn a lot of rocks.

With these few ingredients, hours of exploration are possible. I have tidepooled off the coasts of California, the Bahamas, and of course Bonaire. I'm looking forward to exploring the coastlines of Maine and Oregon. Each coastline is

The long-spined Diadema *sea urchin is a rare sight in the daytime.*

unique, and each type of pool has its own inhabitants. I have found octopi and sea hares and nudibranchs (all shell-less mollusks) and clingfish and eels and crabs. I found my first clingfish on an old bottle on the beach. I had no idea what it was. I carried the little fish home where I could study him more carefully and compare him to illustrations in my books. Of course, I returned him when I was done.

It is also important to return rocks to their original positions so as not to disturb the animals and plants that live on top of and beneath the rocks.

Tidepooling is a wonderful introduction to creatures of the sea but it is far from ideal: the rippled surface of the sea is always between you and the animals. Some creatures— crabs, urchins, starfish, for example—can easily and safely (for them *and* you) be lifted from the water and studied, but some of the most fascinating creatures cannot. The chain moray is not willing to let itself be touched by an alien hand from the world of air, much less actually be lifted from the water. Nudibranchs become lifeless lumps of jelly when removed from their environment. And just try to pick up an octopus!

Avid tidepoolers soon get more than their feet wet, and plunge into the water eager for discovery.

II

Snorkeling

Geared with only mask, fins, and snorkel, I have seen a battle between a triton trumpet and a starfish. I have watched conch as they mated. I have seen sharks feeding on a school of minnows. I have been lucky enough to catch brief glimpses of whales, dolphins, manta rays, and a whale shark. Unencumbered by a tank, a skin diver can move quickly and watch animals in action near the surface of the sea.

Snorkeling is a terrific way for a person to be introduced to undersea life. It opens the sea to children too young for scuba and to adults not prepared for the "rigors" of scuba diving. People with a condition that might prevent them from scuba diving, such as ear or lung problems, can still snorkel. All anyone needs for snorkeling is a mask, a snorkel, and fins—plus an interest in the sea.

In clear water fifteen feet deep or less, even an unskilled breath-holder can enjoy snorkeling, because almost everything can be seen from the surface. One summer I was a counselor at a camp on the island of Antigua, where my favorite activity was fish feeding. Each day I'd lead a group of kids around a bay that had huge mounds of clubby finger coral (which looks just like what it's called), grass flats, and, at the edges, coral rocks. When we began the activity, the fish were terrified of people and disappeared when they saw us. After we had chummed with bread and fish for a week or so, some of the fish, especially the wrasses and small groupers, were hiding less and beginning to feed.

It was interesting to see how each type of environment in that bay had different animal life. The mounds of clubby fin-

ger coral were a mystery to me. Here was a bay with no true corals growing around its rocky edges, but right in the middle of the bay, surrounded by turtle grass, were huge mounds —about twenty-five feet long, eight feet wide, and almost six feet high—of lush, living, feeding clubby finger coral. The mounds also supported other life, some readily visible, like the flowery plume worms, filter-feeding through gill clusters, or the brittle starfish whose arms entwined the inner branches of the coral. When we spread tiny pieces of fish around the coral mounds, some of the shyer inhabitants would be lured out: hermit crabs, carrying their shell-homes around on their backs; spider crabs, with rough shells in colors ranging from red to brown; arrow crabs, with teardrop-shaped bodies and tiny blue claws at the end of long spindly legs. A spiny lobster emerged occasionally from the finger coral mounds, waving its antennae back and forth to seek out the source of the delicious aroma. Unlike Maine lobsters, these animals have no large claws, or any claws at all. Their defense is the sharp spines over their shells and on their antennae. On the antennae, the spines point forward—away from the lobster's body—and predators trying to reach this critter from the front are barred by the waving antennae. Predators trying to get at the lobster from the back are often met by another kind of spine: that of the long-spined sea urchin in whose company the lobster usually lives. Warm-water lobsters do not need claws to protect themselves.

Along with the invertebrates, lots of fish live around the finger coral mounds. As the campers and I approached the mounds, we saw a "halo" of damselfish. The two-inch-long damselfish are quite territorial, so that their spacing from each other is very regular. They all hovered twelve to eighteen inches above the coral, forming the "halo."

As we released bread or fish bits near the coral, schools of wrasses darted in and devoured them. The wrasses are among the most fearless of the little fishes and are usually the first at the food. Each group of wrasses contained numerous yellow females and males, and one or two cigar-shaped "ter-

This young snorkeler has attracted a group of blue tangs with a bit of bread.

Two yellow wrasses investigate a sponge.

minal'' males. The females spawn with groups of the yellow males: an egg-laden female rushes upward, followed by the males, who all fertilize the eggs. The females also spawn with the larger terminal males, but these spawnings are with individuals instead of with groups. If the terminal male dies or disappears, one of the larger females changes into a male and takes over the terminal male's role. This change of gender is described so matter-of-factly in textbooks, but it is so magical!

Occasionally we saw these little wrasses involved in group spawning, but more often they were involved in group *eating*, following us around and gobbling up our food. After only a few days of feeding, these wrasses were ''tame'' enough to nibble bait as we held it in our hands—a good opportunity for the campers to look closely at a fish that otherwise would not stay in place long enough to be studied.

As we left the finger corals, we swam over turtle grass to our next feeding station. With blades six to twelve inches high and three-quarters of an inch wide, this grass is one of very few flowering plants able to live in the sea.

We didn't feed many fish around the grass, but we saw some interesting animals there. One that was fun to examine was a short-spined sea urchin. Camouflaged by a coating of grass, shells, and rubble, these urchins successfully hide from predators—but my sharp-eyed buddies spotted them precisely because of their lush cover. ("Hey! What's that little clump down there?") They'd bring the "clump" up to the surface (keeping it submerged, of course) and gently remove the materials covering the urchin. The kids discovered that the urchin clings to its camouflage by using tube feet, which scientists call pedicellaria; the animal controls them with water pressure. After a snorkeler holds the urchin on his hand for a couple of minutes, he can invert his hand and the urchin will stay attached. We learned that suckers on the ends of the tube feet not only hold the camouflage to the urchin, but also hold the urchin to its rock or pier piling or to the sea floor.

Tube feet allow the West Indian sea egg, a type of sea urchin, to hold on to this divers hand.

Once, as we swam over the grass, we saw a school of fifteen squid, moving slowly, midway between the bottom and
the surface. Gauging the effects of current and gravity, we
dropped pieces of fish so that they floated over to the squid.
Zip! Out went one squid's tentacle and a piece of fish was
grabbed and drawn to its mouth. None of the other squid
seemed interested. Although we couldn't make friends with
them, we enjoyed watching them hover in the water as they
rippled their fins and changed from rust-brown to beige to
mottled to . . . invisible. We would have watched them longer, but we were anxious to reach the next set of feeding stations: coral rocks around the perimeter of the bay.

The rocks were inhabited by long-spined black sea urchins—which we avoided, of course—plus boring urchins,
crabs, shrimp, and a variety of fish. Coneys and graysbys
represented the grouper family in the rocks; both these types
of groupers grow to about a foot long and are very greedy. It
is easy to believe that the saying "stuffed to the gills" was
first used to refer to a coney. These gluttons will continue to
take food until they literally can't fit any more into their bodies. All groupers eat by opening their mouths suddenly and
sucking the food down; coneys will eat until the last piece of
fish actually sticks out from their throats, and even then
they'll try to eat more, even though they're so stuffed they
can't create any suction to draw the food in! Finally acknowl-
edging that they're through with dinner, they sluggishly fall
to the bottom and rest near the protection of a coral out-
cropping. The next day, they're ready to eat again. We had
some fat coneys in that bay!

The feeding habits of the squirrelfish were the biggest surprise to me. Squirrelfish are notoriously shy. Generally, at
the mere approach of a diver or snorkeler they dart into a
crevice in the coral. In our feeding stations that housed
groupers, the groupers took the majority of the food. Here
and there in the bay, though, was a coral formation with
squirrelfish and no groupers. At first we approached these
areas taking special care not to make sudden movements or
kick noisily. We spread out our succulent offerings and
watched quietly as a squirrelfish or two, attracted by the

scent of the fish, emerged timidly from the safety of the coral to snatch the food and dart back. It took them about two weeks to learn that the free meal was safe; soon, they ate our fish right out in the open and waited for more. By the end of the summer, many of the squirrelfish took food from our hands, a special testimony to the campers' gentleness.

The moray eels weren't in the clubby finger coral mounds; they were in the rocks. Although I didn't feel that it was safe for the kids to feed the morays—their control in the water wasn't good enough—I fed them myself. Why? Because we all enjoyed seeing the eels and watching them eat, and because by feeding them I could help dispel some of the fears that my buddies had about eels. After all, what is an eel but a long fish with lots of muscle and very sharp teeth?

The eels we saw most regularly were spotted moray eels. They grew to about three feet long in the protected bay. At first we would come upon them resting in a cavity of a rock, only their heads visible, but as they became accustomed to

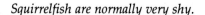

Squirrelfish are normally very shy.

being fed, they looked for us to visit. By the end of the fifth week a couple of the morays would swim up to the surface to greet a snorkeler. Imagine innocently snorkeling along in ten feet of water and suddenly feeling the silky skin of a moray eel as it entwined itself within your arms! That situation was a little unnerving, so we avoided "Moray Castle" for the last week of camp. The eels won't hurt you—they are only looking for fish—but their docility is difficult to believe when they swim up to your mask, teeth gleaming.

Everyone at camp enjoyed watching the animals learn that people could be okay. The kids had the opportunity to see a drastic change in the behavior of these animals—thanks to a little food, some time, and snorkeling equipment.

The bay in Antigua had no coral reef. Most people would have snorkeled it once, and rated it as mediocre to bad. But it did have lots of fish and some corals and fairly large rock formations; it taught us a great deal and we enjoyed it tremendously.

In the summer of 1974 I worked in a dive shop in Islamorada, which is located in the middle of the Florida Keys. The coral reefs off the Keys are several miles offshore; a boat is absolutely necessary to reach them. The waters closer to shore are rated "not worth rating" by most, yet, whenever I had some free time, I snorkeled in those very places that bored everyone else. I especially liked the Gulf sides of Islamorada and Key Largo.

What was there? For miles, the waters were shallower than ten feet. The bottom was either fine sand, grass, or an algae cover. There was little or no coral, but an occasional sponge or a rock or an old tire broke up the contours of the bottom. There were no conch to collect or fish worth spearing. But since I wanted to watch and learn—and not take anything from the area except knowledge—I never got bored.

There is life everywhere in the sea (with some exceptions caused by dredging or pollution). In some places, life is lush and obvious, as in a coral reef or a kelp forest. In other places, the observer must be a little more patient, a little more curious, a little more persistent. I have never been disappointed. There is always something to see.

Off Islamorada I found nudibranchs grazing in the algae. Essentially, a nudibranch is a snail without a shell, a slug. But it is so much more than that! The type I found most commonly, the lettuce sea slug, is ruffled and comes in pastel shades of green, blue, even off-white. The casual observer completely misses this critter because it blends perfectly with the algae or looks like a little sponge or other growth in the sea grass.

I found a lettuce sea slug and gently lifted it up and studied it. It doesn't really look like lettuce; the nickname ''ruffle-backed nudibranch'' describes it much more accurately. Like a snail, its underside is its foot—after a little while in my hand, this nudi attached itself and began to move along my fingers. It had an appealing little face with two projections that looked like the ears on a rabbit. As it cruised along my hand it occasionally lifted its head up as if to get a better view of the surroundings. I placed it back on the algae and made sure it was securely attached before I left it.

The lettuce sea slug is the most common of Caribbean nu-

The lettuce sea slug, or ruffle-backed nudibranch.

dibranchs and also the largest, growing to three inches or even larger. At times I spied other nudis living in the algae, but they were always much smaller—less than one inch. Their beauty makes up for their small size. It's a shame, but they are much too small to be photographed with most underwater camera systems.

Shelled relatives of the nudibranchs also live in the bays off the Keys. The most common ones are the tulips, usually clean (not overgrown or encrusted), in shades of brown and beige, with stripes along the smooth whorls of the shells. The animal in the tulip shell can be colored dark brown, brick-red, or even orange. Upon spotting a tulip shell I always grabbed it immediately. As I raised the shell, the animal would retract inside, blocking the mouth of the shell with its brown horny operculum (the "trap door" of the shell). This behavior is typical of snails; the shell is their protection. What was unusual about the tulip was that almost immediately the muscular foot extended out of the shell again, flailing around for a foothold. No shy snail there! It actually reached around to try to push my hand off its shell.

Once a friend of mine found a pretty tulip shell, which she put into her invertebrate aquarium. By the next morning her six flamingo tongues were six empty flamingo tongue shells, and the tulip was satiated. We learned—the hard way for the flamingo tongues—that tulips are very aggressive carnivores. Their usual prey is shelled mollusks, including other tulip shells, but I found out more about that firsthand.

After my friend's aquarium experience, I stopped grabbing tulip shells and began watching them. One day I was amazed to see a big tulip attacking a smaller one. The two shells were near each other on the bottom. Suddenly the big shell seemed to become aware of the smaller and headed toward it. The little one, realizing its predicament, tried to rush away—not with the usual snail's crawl, but by hooking its operculum in the bottom, vaulting over its foot, hooking its operculum ahead of it, and so forth. It was hopping right along; but the big tulip could hop too, and with a longer stride it easily caught up with the smaller creature. The predator wrapped its body around its prey. At this point I picked

Baitfish in the waters off Bonaire.

them up—the disturbance didn't bother the big tulip at all—and separated them. When the little animal's foot emerged from its shell I was amazed to see a cut in it, made by its attacker, that looked like a two-inch-long, quarter-inch-deep razor slash.

I replaced the combatants separately on the bottom and watched. The chase began again immediately—I guess they could scent each other—and when I left them the big one had easily recaptured the wounded animal. The next day I returned to the spot and found a clean, empty tulip shell. I have found many empty shells, but this is the only one whose history I know.

It's interesting that the best snorkeling experiences I have had have taken place in the least promising spots. On the island of San Salvador in the Bahamas is a dock where people have been cleaning queen conch shells for years. Men go out in small boats and dive for the conchs. They return to the dock, knock a hole in the spire of each shell, cut the animal's muscle away from the shell with a knife blade inserted through the hole, and remove the entire conch. The meat is kept for food, the shell tossed into the sea at the base of the dock. As you can imagine, there are quite a few conch shells around the dock—with holes in them and with encrustations marring the once-smooth pink surfaces of their lips. The shells may not be providing protection for the conchs anymore, but they still support life. While everyone else snorkels out to the reef, why don't we see what we can find around these shells?

At first glance, things look pretty dead here. Conch shells are piled on top of more conch shells. Fine silt covers everything. Let's turn over one of these shells . . . Ah! Here's where all the life is! Exposed to the light, several brittle stars begin to disentangle themselves from each other and seek the comfort of darkness. The arms are all different colors and are either thin, like lizard's tails, or fuzzy or spiny. There are five arms to each nickel-sized body. If you try to pick up one of these squiggly creatures, you'll learn why they're called "brittle stars"—the arms break off at your touch. If there's a puddingwife or a Spanish hogfish around while they are exposed, you'll understand why the starfish head so frantically for cover: If they're slow, those greedy wrasses will gobble them up.

Since the creatures living within and underneath the conch shells live there for protection, we'll be sure to replace each shell so that it's oriented the way we found it. Back goes the brittle stars' shell, and up comes another, under which we might find a cluster of creatures that resemble earthworms. As we disturb their shell, white "fuzz" emerges along the edges of the worms. They are called "bristleworms" because of this fuzz, and the comparison is accurate: they are segmented worms, related to earthworms. They are not brown or ugly, though: they're red or green, and beautiful. In their bristles they have a defense earthworms don't: the white bristles can cause a painful reaction on the skin of a diver or in the mouth of a predator. Like the porcupine on land, the bristleworm is generally left alone. When we turn the bristleworms' shell back over, we'll be careful not to touch the animals.

Here's a shell lying lip down; an eye protrudes from the hole in the spire. An eye? This shell is the home of a small octopus, who peeks out of the hole to watch the world go by. As we turn the shell over and expose the large opening, we can see the octopus's arms squeezing up into the spire of the shell, out of sight. How can we get this appealing little fellow out of his shell? We certainly can't force him—an octopus has too much strength in his arms to be pulled anywhere he doesn't want to go. We'll have to out-think him. We can hold the shell out of the water and see if the octopus will desert his dry home. Since we won't hold the shell out for long, we won't hurt the animal.

Ah! Here comes one arm, sinuous, its underside covered with little suckers that expand and contract separately, seemingly moving in ripple waves as the octopus tentatively feels for the lip of the shell. Then another arm appears, and an eye. He spots us and ducks back into the shell. Finally, discomfort overcoming fear, the little octopus emerges from his home and plops back into the water—onto my waiting hand.

A bristleworm displays its protective bristles.

He spurts a cloud of ink and darts off, a clever trick. The ink takes about the same shape in the water as his body and we are forced to make an instant decision on which "octopus" to follow. Pumping water through his mantle and out his gill hole, he jet-propels his bullet-shaped body down to the grass, where he curls his arms, changes the texture of his body from smooth to rough and his color from gray to mottled, scrunches down on the bottom, and disappears! Oh, he's still there, he just blends in so perfectly he seems to disappear. We can't leave him like this, away from his home and vulnerable, so we place his shell right next to him. He sees it and flows in. Now we can replace his shell in the spot and position in which we found it, and sunbathe with a clear conscience.

Snorkeling has many advantages, and obviously I've had some exceptional underwater experiences without my scuba gear. However, most of those experiences weren't on a coral reef. The luxuriant life on a coral reef invites me to interact with the animals for long periods of time, something I can't do holding my breath. While snorkeling, I can't stay underwater long enough to feed and cuddle an eel. I can't stay down long enough to tickle an anemone. While snorkeling, I can't take a series of photographs, or introduce myself to a scorpionfish, or be cleaned at a cleaner station. To touch the sea, I need to stay under the surface for more than one breath. This is when scuba becomes necessary. Having the skill and the equipment to stay down doesn't guarantee that a person will have the right attitude about touching the sea, though; this became apparent during my early diving experiences.

III

Touch It and It Goes Away

On my first dive, one of the things I was shown was a Christmas tree worm. My instructor pointed to one there on a coral head: two matching "trees," white and delicate-looking. As he pointed at them he moved his finger closer and closer until suddenly — they were gone! All that was left was a little hole in the coral head. I saw another pair, red this time, and swam over to inspect them. Zip! Gone. I realized they were retracting into the holes. What a sense of power this gave the novice Dee! I spent the rest of my dive poking my finger at the poor Christmas trees, watching them disappear.

Later I learned that the animal is a Christmas tree *worm* (*Spirobranchus gigantea*). "A worm?" I thought. "How can a *worm* be so pretty?" (That was my first lesson in how animals are changed in the sea: they lose any repugnance they may have on land and become graceful and beautiful.) The body of the Christmas tree worm is segmented just like the common earthworm's, but divers never see the body because it's permanently encased in a calcareous tube built by the worm. Sometimes the tube is on the outside of a rock and easy to see; other times the tube has been built on living coral, and the coral has grown around the tube, hiding it from view. In either case, this worm cannot crawl around seeking food. Instead, the two little Christmas trees act as filters to remove food from the water. (They also function as gills.) There are other filter-feeding worms with calcareous tubes as well as

37

This worm's feeding apparatus resembles a glorious fountain.

some with noncalcareous, parchmentlike tubes. Some filter-feeding worms are extremely shy, retracting into their tubes so rapidly that only a sharp-eyed diver will see them at all. Others, perhaps because they are less preyed upon, will allow their feathery filters to be touched by a gentle diver. All of them, under stress, retract their "feathers" into the safety of the tube.

A person could spend many dives appreciating the beauty of these worms: red ones and white ones and orange ones and purple ones. Double Christmas-tree–shaped filters, feather-duster–shaped filters, puckered-lip–shaped filters. Some live alone; some live in groups. I have seen them everywhere I have dived. I didn't expect to see them in the cold waters off California but they were there—and in shades of blue I had never seen in the Caribbean.

The filter-feeding worms were the first animals of a category I began to call "Touch It and It Goes Away." Many un-

The twin filter-feeders of the Christmas tree worm.

dersea creatures expand when they are feeding but retract when danger threatens. These, the "Touch It and It Goes Away" animals, are very delicate; by retracting in response to stress or danger, they save themselves from injury. When we touch them gently, we cause no harm and are able to see their reactions. We just have to be careful not to overdo.

Touch It and It Goes Away animals fall into the phyla of annelids (the worms), coelenterates (corals and anemones), and a few of the mollusks.

The coelenterates are the polyp animals, including the stony corals, the soft corals, and the sea anemones. A polyp is a tubelike organism. Coral is an interdependent colony of polyps. To many divers, the stony corals look like peculiar rocks. But these colorful, rocky formations are only the calcareous skeletons within which the soft animals live. In the Caribbean, corals are colored mostly in muted greens and browns, with an occasional reddish tint. Most stony corals have specially adapted algae living in them: the algae, which is a plant, uses sunlight and carbon dioxide for the process of photosynthesis, and produces oxygen as a waste product. The corals, which are animals, use oxygen for their metabolisms and excrete carbon dioxide as a waste product. My description of this process is obviously simplified, but it helps explain why algae and corals live together so well. It also explains why corals do not live in deep water: below two hundred feet or so, algae receives insufficient light for photosynthesis. The partnership of algae and corals also explains the colors of the corals: they take on the shades of the algae living within them.

Some corals, like brain coral, grow in very sturdy, solid shapes and are capable of building huge hemispheres above the bottom. Other corals are more fragile: staghorn coral grows slender arms, pillar coral grows in towers, and pencil coral grows in clumps. When the stony corals feed—most feed at night but a few are diurnal—their polyps extend over the stony skeleton like tiny fleshy fingers. If a diver creates a disturbance near the extended polyps of a coral, the polyps disappear. Another example of Touch It and It Goes Away.

More obvious for Touch It and It Goes Away than the

Coral polyps, extended for feeding.

stony corals are the soft corals, also called gorgonians. Novice divers often register the soft corals as plants because they sway back and forth in any surge just as plants do in a breeze; they frequently appear to have a trunk, which seems to be rooted in the bottom, and extend upward in branches, and when feeding they even seem to bloom. There are very few flowering plants in the sea, though—actually, there are very few plants at all in the sea—and a close inspection of a soft coral will reveal that the "blossoms" are actually tiny polyps, extended, seeking sustenance from the passing water. My first experience with one of these was accidental: I brushed against it as I swam along. Suddenly the entire appearance of the gorgonian changed. The fuzzy tree became a smooth tree! I spotted another fuzzy tree and this time deliberately ran one finger along a branch—the fuzz disappeared from the branch, then in a chain reaction the fuzz disappeared all over the tree. I looked very carefully before touching the next gorgonian and realized that the fuzz was composed of tiny eight-fingered polyps. When the coral was disturbed, the fingers of

each polyp closed up—the reverse of a flower blooming—
and if the coral was seriously disturbed, the entire polyp
could retract into a pore on the branch. Touch It and It Goes
Away!

Along with the hard and soft corals, the other coelenter-
ate members of the Touch It and It Goes Away category are
the anemones. If the soft corals are the "trees" and the
stony corals the "rocks" of the undersea garden, then the
anemones are surely the exotic "flowers." The giant Caribbean anemone *(Condylactis gigantea)* has three-inch long,
quarter-inch thick, polyp-tentacle "petals," tipped in
pink, or purple, or green, or yellow. Another common Caribbean anemone is the ringed anemone *(Bartholomea annu-
lata),* with very thin, clear tentacles that have white lines
spiraling along their lengths. The less common forked
anemone *(Lebrunia danae)* has some arms that are brown
and others that are milky white. Other anemones come in
still other patterns: the white-dot anemone *(Heteractis lu-
cida)* has white dots decorating its otherwise clear tentacles; the "sticky" anemone *(Stoichactis helianthus)* has
very short, nubby tentacles; and the orange ball anemone
(Psuedocorynactis carribbeorum), which only opens up at
night, has a bright orange ball at the tip of each of the clear
tentacles that surround its mouth.

Unlike the corals, which are permanently attached to
the substrate, each anemone is attached by its foot. It is
able to move around but generally stays in one place.
(There are other differences between corals and anemones,
of course. For example, the corals are colonies of animals,
the anemones single animals; corals have skeletons, anemones do not.) Anemones usually attach themselves to a
protected spot—inside a dead conch shell, underneath a
rock, within a hole in a coral head—and extend their tentacles to feed outside their protected area. All of these anemones, when stressed, pull their tentacles back into the protected area, or, as in the case of the orange ball anemone,
simply curl their tentacles up within their body cavities.
Touch It and It Goes Away.

The common Caribbean sea anemone.

The last of the Touch It and It Goes Away animals that I discovered in my days of power-play diving were the mollusks; including the cowries as well as the trivia, cyphoma, and simnia shells. These animals share the common characteristic of very clean, and in most cases, a highly polished shell. The shell of each mollusk extends its mantle (the thin sheet of tissue that secretes the material of the limy shell) in order to completely cover it, camouflage it and, protect it from encrustation.

Of these mollusks, the one most seen by divers and even snorkelers is the flamingo tongue shell. They are usually found in pairs attached to the branches of various soft corals. They look so pretty! The flamingo tongues are about an inch long, creamy yellow in color with contrasting black rings all along the shell. "Aha!" many divers think, "I want to bring that one home!" and they pluck a shell from the coral. The pretty black circles disappear, leaving a rather plain cream-colored shell. The attractive pattern seen on the flamingo tongue shell is actually the fleshy mantle of the living animal, enveloping the shell. When the animal is disturbed, the mantle is withdrawn into the protection of the shell. Touch It and It Goes Away.

Cowries are seen by divers much less frequently than flamingo tongues because their habits are more reclusive and also because their camouflage is better. Instead of having a brightly colored mantle, cowries glove their shells in a warty, dull mantle that makes them look like uninteresting sponges. In addition to being well camouflaged, cowries spend their daylight hours hidden under coral heads or deep in rock ledges where divers cannot even see them. Divers rarely get the opportunity to recognize cowries in the daytime.

At night, though, cowries come out to forage. I recognized my first living cowrie by inadvertently bumping into a brown "sponge" on a night dive. The "sponge" split in half (the animal's mantle encloses its shell hemispherically) and withdrew like something in a science fiction movie, and the glossy shell was left bared for me to see. Touch It and It Goes Away.

The brilliant mantle of a flamingo tongue mollusk. When the animal is disturbed, the mantle retracts taking the spots along with it.

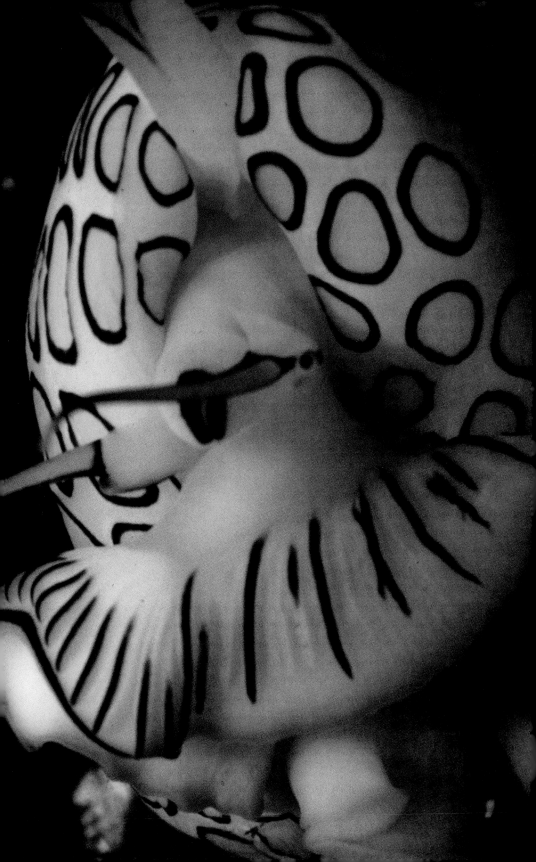

Because the cowrie's shell is so well protected, it is coveted by collectors. It is a shame that most of the collectors never see how perfectly the living cowrie adapts to its environment.

So, my first lesson in undersea touching was that everything goes away! Anemones, coral polyps, mollusk mantles. I thought, "There has to be more to touching underwater than watching animals disappear." I was right, although it took me nearly five years to become more gentle and thus to discover how best to touch the sea.

IV

The Invertebrate "Rogues' Gallery"

". . . and have a good dive. If anyone would like to dive with me for a personalized tour of the reef, just let me know."

"Oh, Dee, Rheta and I would love to dive with you!"

"Aha!" I thought. "This is my chance!" Jeanne and Rheta had been using thick wetsuit gloves all week, which was certainly overkill for the Caribbean. "Terrific, Jeanne, but you'll have to leave those gloves on the boat!"

"We'd love to leave the gloves. We just didn't know what was safe to touch. We've been afraid to take the gloves off."

Touch the Sea was thus begun in September 1979.

I had watched so many people dive encased from fingertips to toes in wetsuits or leotards and tights, but it had never crossed my mind that many of them were covered up because of fear. Then I remembered the "Marine Environment" lecture of so many scuba courses. (I remembered it because I used to give it.)

Slide #1: Photo of shark. (Beginning a slide show with a shark always got everyone's attention.) "This is a shark. Sharks have sharp teeth and can grow to be very big. Since sharks are potentially dangerous and very unpredictable, you should leave the water if you see one."

Slide #2: Photo of barracuda. "This is a barracuda. 'Cudas get as big as six feet long, and they like to strike at shiny objects, so take off your jewelry or cover it up when you dive. You will see lots of barracudas in the Florida Keys, and no

47

one has ever been bitten, but I read in a book once about a barracuda that . . ."

Slide #3: "This is a moray eel. Eels have long sharp teeth, which grow along their jawbones like ours, but they also have an extra row of even longer teeth, growing right down the middle of the roofs of their mouths. When they bite down on something, all those teeth give them an excellent grip, as you can imagine! Moray eels live in crevices in the coral, so never stick your hand into a hole after a lobster or anything without checking for an eel first."

Slide #4: "Can you see the fish in this slide? There in the middle? It is a scorpionfish, one of the Caribbean's venomous fish. The scorpionfish lies quietly and invisibly on the bottom, and when a small fish investigates the "algae-covered rock," the fish is sucked in for dinner. Divers don't have to worry about being eaten by scorpionfish, but if they should inadvertently place a hand on the fish, it would erect its dorsal spines and zap the hand. Scorpionfish venom isn't likely to be fatal but it is quite painful."

Slide #5: "This is fire coral, and boy, can that stuff sting! The more I dive the more sensitive I become to it, too, so it is always good to be covered up when you dive—either with a wetsuit, or long pants and a sweatshirt."

Slide #6: "This is a bristleworm. It is related to our common earthworm, but you see that white fuzz on its sides? That is the stuff that gets you. It sticks in your skin and feels like little fiberglass bristles. There is really no first aid for the pain, either, though some people favor trying to remove the bristles with Scotch tape."

Slide #7: "This is a long-spined sea urchin, called by scientists *Diadema antillarum*, 'crown of the Antilles.' If a person falls on or steps on or places a hand on those sharp spines, the spines will penetrate the skin and break off. They are very thin and impossible to remove; luckily, the body dissolves them—eventually. According to some scientists, the spines of *Diadema* contain a toxin which . . ."

* * *

Fire coral with extended polyps.

Who wouldn't be afraid to venture into a sea filled with malevolent creatures that sting, stab, cut, bite, envenomate, or perform some combination of those nasty things? No wonder many divers are out there to prove their bravery. The sea has been presented as one huge challenge to anyone foolhardy or courageous enough to enter it.

I have toyed with the idea of calling this section *There Is No Such Thing as a Dangerous Marine Animal.* The statement is an exaggeration, of course, but, like slides of sharks, it would get people's attention. In the next few chapters I will try to explain and defend the so-called "dangerous marine animals."

Under what circumstances would an animal be likely to hurt a person? One could crash into an animal, such as a coral, and be damaged by the collision itself. An animal might defend itself against a diver who is perceived as a threat. Or an animal might mistake a diver—or part of a diver—for food.

The *knowledge* that one should avoid crashing into things underwater is as basic as the knowledge that one should not walk into trees or lampposts. The *ability* to avoid crashing into things comes from good training in buoyancy control, which every diver should glean from a scuba certification course.

To avoid stressing an animal or being mistaken for its food, you simply must know about the animal and its habits. The more you know about an animal, the more comfortable you will be about interacting with it. The more comfortable you feel with it, the more often you will interact with it. The more often you interact with it, the more you will know about it. Enter this not-at-all-vicious circle of interaction anywhere—you can only come out ahead.

So, let's begin with the "stingers." A classic example is fire coral. (Though not actually a true coral, it looks much like one.) Fire coral grows in fingery-flame shapes reaching toward the surface of the water, or in flat rippled plates, or as encrustations on dead corals. It derives its name not from its color, which is a golden-mustard hue, but from its effect: brush against fire coral with your arms and you will feel as if you have been brushed with flame. The burning feeling lasts up to a few hours, often leaving a raised, angry-red welt.

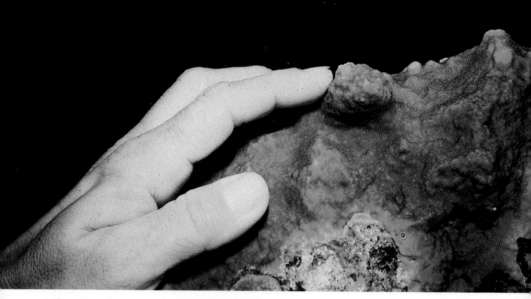

When its polyps are retracted, fire coral is safe to touch.

Does fire coral leap up off the bottom to "burn" divers? It only seems that way. Divers with good buoyancy control and a good sense of where they are in the water can easily avoid being zapped by fire coral. Divers who are less sure of their skills or who are otherwise occupied can wear protective clothing.

That's as much as most people want to learn about fire coral. Don't go away, though! This stuff deserves more attention. Look closely at some fire coral, especially around the edges. (Or, check the photographs.) See the fuzz? Those are the polyps. Some polyps are adapted for feeding, others for stinging. Within each stinging arm are nematocysts, the fire coral's built-in "harpoons." When touched, the nematocyst is triggered; it fires a stinger and envenomates the offender. How fragile that "fuzz" looks, though—can anything inside something that frail actually penetrate human skin? Well, yes, it certainly can—almost anywhere on the body. But consider the skin on a person's hands. It is thick. If you are beginning to think that fire coral nematocysts won't penetrate people's hands, you are right. Really. Fire coral cannot sting the palms of people's hands.

The macho part of me loves to point out fire coral to my buddies, signal that it burns and is not to be touched, and then repeat the signal after *holding* the fire coral (barehanded, of course!) to make sure they know what I am referring to. Upon surfacing someone always remembers to ask, "Wasn't that fire coral you were pointing out? How can you touch it without getting burned?" "I'm so glad you asked . . ." and I explain to them what I have just explained to you.

There is something else about fire coral worth noticing: once it has been touched, the polyps retract into the skeleton. And since the polyps hold the nematocysts, once the polyps retract it is safe to touch the fire coral with any part of your anatomy, thick-skinned or not, until the polyps extend again.

Now the thought of fire coral should not be as disconcerting as it might have been before you read this section. Just don't get as relaxed with it as did a man I once dived with: he saw an old whiskey bottle lying on the bottom, and in a spirit of playfulness he took his regulator out of his mouth and put the bottle there, as if drinking. He had not noticed that the bottle was encrusted with fire coral, but his lips reminded him of that fact for the next few hours. Ouch.

Another group of "stingers" that many people are afraid to touch is the anemones, the "flowers" of the sea. My first experience with an anemone was as a Touch It and It Goes Away animal, because when I reached for it, it shrank from me. Anemones use their tentacles to feed, and, like those of fire coral, their tentacles contain nematocysts. The anemones' nematocysts protect the anemones and paralyze their prey.

One day while watching a Caribbean anemone (*Condylactis gigantea*) I saw a little purple-and-clear shrimp hopping around on the tentacles. I wanted to look more closely at the shrimp, so I inserted my fingers behind it and encouraged it to hop onto my hand. As I did that, I remembered that anemones are supposed to sting—but although the tentacles were sticking to my fingers I didn't feel any stinging at all. As the

Above, a spotted cleaner shrimp lives in this common Caribbean anemone.

Below, the nematocysts of the common Caribbean anemone are too weak to sting through most human skin.

tentacles of the common Caribbean anemone stuck to my fingers I slowly pushed my hand farther into the anemone, until tentacles captured each of my fingers as well as the palm and back of my hand. (The skin on the back of the hand is not thick; apparently this anemone's nematocysts are too weak to sting through most human skin.) When I tried to remove my hand from the animal the tentacles felt like weak Scotch tape as they came unstuck—so I called these anemones the "Scotch tape" animals. On later dives I became fascinated with watching anemones try to "eat" my hand.

Some anemones have more potent nematocysts than others, and I have been stung on the back of my hand but never on the palm. The stickiest anemone I have found so far is the one I call the "sticky" anemone *(Stoichactis helianthus)*. When I taught diving in the Bahamas, I always showed divers a sticky anemone on their first dives and showed them how, when they touched the nubby tentacles with their fingers, the tentacles would stick to them. Imagine my surprise when I read in Patrick Colin's *Caribbean Reef Invertebrates and Plants:* ". . . a sharp sting is felt if uncalloused skin . . . is applied to the anemone . . . the skin may blister in the area of contact." On reflection, though, I realized that my students and I had only been touching the anemone with the thick skin on the palms of our hands. Since then, I have touched the sticky anemone with the back of my hands and although I felt the prickly stinging and my hand itched a bit, I have never blistered. Still, it is best to experiment in touching the sea with fingers instead of more sensitive skin.

An animal that reputedly feeds on anemones and that I have observed eating fire coral and staghorn coral is the bristleworm, our two-to-twelve-inch-long marine relative of the earthworm. (In the chapter on snorkeling, we encountered bristleworms living in conch shells.) The bristleworm is much more beautiful than the earthworm, though. The common bristleworm *(Hermodice carunculata)* has a reddish body with bright white bristles.

When the bristleworm goes about its business, the bristles remain in modest clusters at the sides of each of its body segments. When the animal feels the need to defend itself, it

Stoichactis helianthus, *the ''sticky'' anemone.*

Double trouble: a bristleworm eating fire coral.

causes the bristles to become fuzzy—much more noticeable—
and, if contact is made with the bristleworm, the bristles de-
tach and embed in the intruder, painfully. Most animals (including divers) only have one bad experience with a bristleworm; one is quite enough to teach even the dullest pupils to
stay away.

Every so often a gentle diver who is unaware of the
bristleworm's bad reputation will innocently handle one—
and remain unscathed. Why? Because the animal uses its
bristles for defense, and a gentle diver does not threaten it.
On one dive I had been feeding fish and my hands were fishy
from the food. I saw a bristleworm moving along the bottom,
lured into the open by the smell of fish. I placed my hand before the bristleworm. The residue of fish seemed to excite it;
without hesitation it climbed onto my palm and began to explore. Then its mouth opened and I felt a rasping sensation
(not unlike a cat's tongue) on my skin. The bristleworm was
tasting my hand! In all, it tasted one finger twice, my palm
twice, and my thumb once before giving up and crawling
back onto the bottom. Every so often, the sea touches back.

Bristleworms don't taste people very often, but some
types of sea urchins regularly "touch the divers"—or anything else they come in contact with—with their tube feet. I
had heard of sea urchins well before I began to scuba dive.
They are the "pincushions of the sea," the dangerous marine animals most likely to zap a diver or even a swimmer,
the villains of rocky seashores throughout the Caribbean. On
my first scuba dive, in Florida's Pennekamp Park, I realized
why they are called "urchins": under every ledge (where I,
novice that I was, expected to find lobsters) were tens, no,
hundreds of long-spined black sea urchins. "Phooey on
those guys!" I thought. "What good could they possibly do?"

It was years later when I finally learned the answer to that
question: urchins eat algae. Algae and coral compete for the
same niches on the reef. Thus, reefs with no urchins become
reefs with lots of algae and not much coral. There is a reef in
the Florida Keys where divers for years have been chopping
up urchins to feed the fish. As the density of the urchin pop-
ulation has fallen, the algae has flourished.

If for no other reason, urchins should be welcomed because they help guard the reef.

But must they have such obnoxious, long, dangerous spines to guard their reefs? Absolutely—because to many reef fish, the sea urchin is a delicacy. That is why, during the daytime, the urchins huddle under ledges, protecting their vulnerable short-spined bottom surfaces, where their mouths are. Some triggerfish will actually push over small rocks to devour the urchins beneath them, or create a water current to try to "blow" an urchin over, exposing the underside. Once when I picked up an urchin by one of its long spines to show it to another diver, a Spanish hogfish shot up under my hand and got hold of the urchin before I had time to say *Diadema antillarum.* All I had left to show the diver was the urchin's test (shell), spines still attached, but bottom and contents—and life—gone.

The spines of *Diadema* have directioned barbs. If you run your finger along the spine from thick end to thin, it feels very smooth; stroking in the opposite direction is almost impossible, because of the barbs. The spine's barbs create resistance going *into* flesh, and no resistance coming *out.* The urchin, even in the construction of its spines, tries not to hurt us, and still we malign the poor thing.

After I understood enough about the anatomy of a long-spined sea urchin to know that the spines on its oral surface were short, I realized that it could be safe to hold one. Step one: find an urchin that is not under a ledge or otherwise protected. Step two: grab the urchin by one of its long spines and lift it off the bottom. Step three: place a hand *under* the urchin. Step four: drop urchin on hand. I have found this procedure to be effective *most* of the time. After cuddling urchins in this way, I find myself eyeing them benevolently and protecting them from would-be urchin hunters. Poor mistreated creatures!

The bad reputation of the long-spined urchin usually extends to its short-spined relatives, and fear again prevents people from relaxing and touching the sea. While *Diadema* proudly displays its long spines, the short-spined urchins, as

my campers discovered, frequently camouflage themselves with bits of sea grass, shells, or coral rubble.

The first time I found one of these short-spined urchins (it happened to be a West Indian sea egg, *Tripneustes esculentus,* a pretty animal with a black body and white spines), I placed it on my hand and carried it to the other divers, pointing out the camouflage. I turned my hand upside down, and the silly thing defied gravity and stayed right in place. Its tiny pedicellaria (tube feet), which the urchins control by hydraulic power and which they use for gripping, were hard at work. When we looked closely at the urchin we could see the tube feet, stretching but gripping tenaciously as I pulled the animal away from my hand. I rocked the creature back and forth until it detached, then passed it around the group of divers so that part of the sea could touch each one of us. Holding an urchin this way is fun (as long as you don't try to shake hands).

If you do want to shake hands, the "rogue" to try it with is the octopus, who has eight arms to offer. This intelligent relative of the snail used to have a terrible reputation, which wasn't helped by Jules Verne's *20,000 Leagues Under the Sea,* but thanks to more educational films the octopus's shy character is better understood now. Still, too often there are reactions of "It will bite!" and "Yucch! Slimy!" when octopi are mentioned.

Octopi *are* capable of biting. Each octopus has a horny beak somewhat like a parrot's beak, and each octopus has a venom apparatus. The prey is caught, envenomated, and eaten at leisure. One of the most potent octopi is the Australian blue-ringed octopus *(Octopus maculosus),* whose venom can kill a person in only a few minutes. On the other hand, no Caribbean octopus has venom even nearly as potent—and I have never heard of any octopus of the Caribbean that has ever attempted to bite anyone.

Once, on a night dive, my buddy and I found a beautiful long-armed brown octopus with white spots *(Octopus macro-*

The author balances a long-spined urchin on her bare hand. Photo by Tom Steinmetz.

pus) on an interface of sand and grass. We followed it and stroked it until it backed up to a rock, spread its arms down, and scooted underneath the sand. In about ten seconds it was gone. Completely buried. I've never read about such behavior or seen it since.

Octopi are rewarding to touch. They can change the apparent texture of their skins from smooth to warty, but when touched they always feel smooth without being slimy (like many underwater creatures, including fish and eels). When the small suckers on an octopus's arm explore my fingers, the independent action of each sucker is what surprises me; it feels like many tiny animals instead of only one very talented octopus.

During a dive I once noticed a tin can on the bottom. As I swam closer I realized it was one of a number of articles—stones, shells, et cetera—marking and blocking the doorway of an octopus's home. With a bare hand, I stroked one of the exposed arms of the octopus, which it then wrapped around my finger to pull my hand inside its den. The octopus drew

This octopus was photographed on a night dive.

my hand in until I could feel the other end of its den, then held my hand in position with several of its arms. Each time I gently tried to pull myself free, the octopus tightened its grip. At the time, I couldn't figure out why it was holding me there, but now I theorize that that octopus saw me as a perfectly qualified doorway-blocker. What better protector for the doorway of his den (from the octopus's point of view, anyway) than a huge, noisy, obedient diver?

Unfortunately for the octopus, however, this diver got cold and became less obedient. I tried to remove my hand. Ha! The octopus was holding on to me and its house at the same time— all that suction was a powerful force to reckon with. I might have been able to pull free by sheer force, but would probably have hurt the animal. Finally I squeezed my other, gloved hand into the den. The animal's attention was divided by the intruder and I was able to reclaim my entrapped arm. Just barely: my hand was covered with tiny red spots where the suckers had been working. Transient souvenirs of the interaction, the little spots were gone the next day.

Finding an octopus's home—detecting the animal by the characteristic decorations at its doorway—always makes me proud of my skills of observation. Except for hand-holding, an octopus at home is not likely to interact with a diver. And it is darned unlikely to come out of its house to say hello.

One time, though, *Skin Diver Magazine* photographer Geri Murphy and I found a little octopus living in an abandoned conch shell. We successfully encouraged him out of his home; in very little time he snuggled down onto my hand and Geri was able to photograph us to her heart's content (with three camera systems). By the end of the shooting session I could stroke the little octopus without frightening him, and he was reaching out a tentative arm or two to explore my face around my mouthpiece. In a burst of bravado I took off my mask and the little fellow walked all over my face; then I replaced my mask, cleared it, reunited the octopus with his conch shell, and put the shell back where Geri and I had found it. (I like to imagine the octopus discussing the experience that night with a friend over a beer: "I had the chance to touch a human today; it was smooth, but it wasn't slimy!")

Fire coral, anemones, bristleworms, sea urchins, octopi: they comprise an Undersea Invertebrate Rogues' Gallery. During years of interacting with these animals I have found out for myself how benign they can really be. The more one learns, the more natural it becomes to touch the sea.

The author cavorting with an uninhibited octopus. Photo by Geri Murphy.

V

The Vertebrate "Rogues' Gallery"

Every Invertebrate Rogues' Gallery must be accompanied by a Vertebrate Rogues' Gallery, right?

Because of its reputation, the logical first nominee would be the shark. Sharks got a lot of media coverage in the 1970s, with *Blue Water, White Death,* the documentary that probably began the fad; *Jaws,* the film that should have ended it; and Jacques Cousteau and *National Geographic* specials on television. As a matter of fact, I have seen hundreds more sharks on television than I ever have in the sea!

Why? Essentially, because a non-spearfishing sport diver is not likely to see sharks. Perhaps the sharks see the divers first and flee, or perhaps there are so few sharks that as long as divers do nothing to attract them, the odds of their paths crossing are slim. We understand sharks so poorly that the reasons for their behavior are unclear. However, it is a clear fact that there is very little likelihood of a shark attack on a diver.

That fact didn't mean much to me the first time I saw a shark, though. I was a new diver and my buddy was almost as inexperienced as I. We had taken his open motorboat out to Fowey Light, a popular spot for Miami divers: it is close, fairly shallow (twenty feet to sixty feet), and has good coral growth for Miami. Peter and I were cruising along, sightseeing. I looked ahead to see a hammerhead shark swimming directly toward us, only ten feet away. Before I could summon my muscles to any action, the shark veered off and swam out of sight. Pete and I walked on water to the boat and got there about a year later. (Our eyeballs kept sticking to the glass on our masks as we peered around looking for the shark . . .) We agreed we were finished diving for the day and returned to the dive shop with our story.

"Carlos!" I shouted as I entered the shop. "We saw a shark today! It was—"

"A hammerhead?" he interrupted.

"Yes! It was about—"

"Seven feet long?"

"Yesss . . ." (cautiously)

"At Fowey Light? About one in the afternoon?"

My gaping mouth prompted him to explain.

"That shark always cruises around Fowey in the early afternoon!"

So much for our unique shark adventure! Oh well.

Unique or not, seeing a shark causes distinct emotional reactions. My first shark sightings were shadowed by my fear of the huge, toothy creatures of ill repute. They sure look

Barbels under the mouth and a tail fin with a long top lobe characterize the nurse shark.

big in the water! I guess my feelings about sharks began to change after a discussion I had one day with filmmaker Stan Waterman. Later that same day, a hammerhead passed within eight feet of me and in my head I could hear Stan's mellifluous voice: "What a magnificent animal!" Concentrating on the beauty helps me forget my fear.

Sharks can be pretty elusive when people are looking for them, as any viewer of the shark documentary *Blue Water, White Death* knows. When I was working on San Salvador in the Bahamas, filmmaker Jack McKenny wanted to film a shark. On a day with no guests diving, we anchored a boat over a point where the bottom dropped off at a site where a shark was occasionally seen. From the surface our captain dropped fish and beef blood, while underwater Jack speared a parrotfish and attached it to a line about thirty feet below the boat. Five of us, armed with cameras and sticks, huddled on the bottom watching ceaselessly for approaching predators. Nothing. After a while, Jack speared another fish so more vibrations of struggling and bleeding would enter the water, and that was when the last of my enthusiasm for the project was lost. In the end, the only fish to show up was a small barracuda that was too timid to touch the bait.

I have managed to have two "unsolicited" interactions with sharks. The first was quite frightening: I was dive guiding in San Salvador at the time, and several divers on my boat were taking down raw—and rather rotten—fish to feed the groupers. When most of the divers had left the water (two were left hanging on lines at ten feet doing safety decompressions), I positioned myself in the water near the anchor to feed the groupers the rest of my coffee can of fish. For some reason I looked off into the distance and saw a six-foot shark casually swimming by. I could tell it wasn't a nurse shark or a hammerhead, but that was all I knew. I replaced the top on the coffee can and watched the shark swim gracefully and calmly. Since the groupers didn't seem nervous, I resumed feeding them. This time the shark changed direction and headed toward me. Suddenly the groupers all disappeared into crevices in coral heads, and there I was, too big to hide with the groupers, with no gloves, no stick, and holding a can of ripe fish.

As the shark approached, I stuck out my leg, presenting the flat part of my fin in the hope that it would be too big for the shark to fit his mouth around. He actually *bumped* the fin (sending my heart into my mouth where it seriously threatened to interfere with my breathing . . .) then turned around and headed toward me again. I thought, "He wants the fish? Okay, he can *have* the fish!" and I dumped the can's contents into the water. Too late I realized that for a six-foot shark, fish cut up for grouper consumption was only an hors d'oeuvre.

In desperation I shoved the coffee can away from me. The shark, without hesitation, opened his mouth and ate the metal coffee can, earning the name "Coffee." While he was occupied, I kicked up to the boat, thankful that my dive was conservative and required no decompression. I casually climbed the ladder with my fins on, and stood on that wonderful, safe, dry deck. On my next few dives I spent a lot of time looking behind me. I didn't feed the fish for a couple of weeks, and I began to carry a stick—a two-foot aluminum pipe with a bicycle handlebar grip—just in case. That particular "magnificent animal" had really gotten my attention!

One day, a mere three months later, I actually used that stick. It was the first day of the diving week; the group on my boat was composed mainly of novice divers. They were all back in the boat, the group leaders (Paul and Sandy Reynolds of Buckeye Divers in Cleveland) were recovering from the checkout dives by hovering at ten feet, and I was contemplating the anchor's hold in the sand at forty feet. When I looked ahead, a five-foot shark was coming straight at me from about three feet away. I held out my trusty stick, the shark swam into it, and I deflected her around me. Paul and Sandy came down to join the adventure. The shark repeatedly swam wide ovals around us, never varying her path or speed. She didn't seem hungry, or aggressive, or excited— just curious. Finally, overcome with frustration at not having even one camera on the boat, Paul, Sandy, and I left the still-circling shark and surfaced.

The next day Paul suggested we return to Vicky's Reef, where we had seen the shark. We didn't really expect to find her there, but off we went, cameras in hand this time.

The first two divers were no sooner in the water, snorkeling to wait for the rest, when they shouted, "There's the shark!" Paul and I leaped into the water and zipped down to the bottom. The shark swam along the reef. We took pictures. She swam along the dropoff. We took pictures. She swam over the sand. We took pictures. To this day I've never had a more cooperative underwater model. We nicknamed her Cheryl Tiegs.

We dived Vicky's Reef twice more to shoot pictures of Cheryl Tiegs Shark, but by then the other divers were reacting with yawns. "What? Dive with the shark *again?* Boring!" So those were our only dives with Cheryl.

Almost two years later I displayed a print of a photo of Cheryl Tiegs at Planet Ocean's Bounty of the Sea Festival in Miami. Marine biologist Sonny Gruber of the University of Miami saw the print (and studied my other shots of Cheryl) and identified her as a female reef shark. She appeared to have some obstruction in her throat and she must have been hungry—Sonny said her ventral side should have been rounded out but it was concave. There we had been, playing with a sick shark. . . .

"Cheryl Tiegs" shark. A remora fish can be seen underneath.

As far as touching the sea goes, sharks aren't your best candidates. No one knows much about them, and because they are so potentially dangerous, most divers shouldn't seek them out. Even the Taylors (Ron and Valerie, of *Blue Water, White Death* and a host of other shark films) have resorted to chain mail armor for some of their interactions with sharks. There is plenty of excitement in the sea with much less risk.

Closely related to the sharks—also cartilaginous, as opposed to bony, fish—are the rays. The stingray has a bad reputation, because at the base of its tail it carries one, or, in the case of the spotted eagle ray, several, venomous spines. I suppose a wader stepping on a ray could get the worst of those spines, but I've never heard of a diver or snorkeler being injured by a stingray. It has been my experience that, when approached by a diver low on the bottom, the Southern stingray (muted gray on top, light-colored underneath) will often stay in place and allow itself to be stroked along its "wings." They feel wonderful, as smooth as wet chamois. Even when frightened, these animals don't respond aggressively as long as they can swim away.

The best testimony I can give to the Southern stingray's good temper is an interaction I had with one while diving off Grand Turk. The ray was lying buried in the sand. I crept up to it and stroked it, all along a wing, under its "chin," and even on its "nose." Then I noticed something protruding from one wing. A hook. A three-pronged snag hook. Tentatively, I tried to pull the hook out, with no luck. I felt gently under the ray's wing to see if the hook went all the way through, which it didn't. I swam back to the boat, got a wirecutter, returned to the ray, and tried to cut the hook where it emerged from the ray's wing. The tool was rusty and dull, and as I struggled to cut the hook the ray was disturbed. It began to swim. I hung on. It would have been perfectly understandable if that ray had zapped me with its spine—I was within range, and I was probably causing it pain—but it didn't. At last the hook pulled free. The ray, newly unencumbered, settled to the bottom and went back to sleep. I, Androcles of Grand Turk, returned to the boat

A Southern stingray.

with the snag hook. If only fishing hooks and lines were instantly biodegradable!

Not two hours after I wrote about the "Androcles incident" I was diving *Mi Cas,* my favorite reef. I used most of a roll of film photographing Popcorn the eel when Oliver Twist, my peacock flounder friend (you'll read more about Popcorn and Oliver in later chapters), cruised by. He came to rest on a bed of thick algae in his dark color phase. Glad to be able to photograph those colors, I finished the roll and put the camera down. Oliver glided seaward, and as I watched him swim over the dropoff, my eye was caught by a spotted eagle ray!

In as unthreatening a way as I could, I swam toward the ray. It soared along, a six-inch remora swimming just beneath it. I swam parallel with the ray, about fifteen feet away from it, admiring its grace as it winged into the blue water. Finally, I headed back to the reef—and saw two more spotted eagle rays "flying" below me. I dropped down to try to accompany them. They let me get very close—almost within arm's reach of them—and I was able to see, better than I ever had before, the pattern of rings on their backs and their smooth white undersides, broken only by the creases of their mouths and gill slits. I noticed that each had several barbs protruding from the base of its tail and I remember thinking that the sight reinforced what I had written that morning. All this time the rays swam so slowly that I had no trouble at all keeping up with them. I began to try to get a bit closer, and they accelerated slightly. Not wanting them to feel that I was a pursuer, I slowed down and headed back up to the dropoff.

As I took one last look in the direction the rays had gone, I saw the first one, quite a bit ahead of me, angle around and swim back toward me. I hovered in the water as the ray, remora faithfully swimming underneath, approached; then I began swimming back to my camera. The ray stayed with me, keeping a comfortable fifteen feet away. I flapped my arms slowly, in time with its wings, and it moved in a bit closer. We kept company until I reached my camera and stopped swimming; I watched the ray leave the reef's edge and head into the blue water.

Did that lovely creature of the sea turn around to accompany ungainly, noisy *me?* The incident could have been mere coincidence. I prefer to believe, though, that an unthreatening intruder into the sea can occasionally be fortunate enough to have special interactions of this type.

Number three in the Vertebrate Rogues' Gallery of the Caribbean is probably the barracuda. Barracudas are smaller than sharks and more predictable. At any rate, some of the actions of 'cudas are easy to understand. For example, they are attracted by anything light-colored or shiny. One "game" to play with barracudas is to hover near the surface and drop a dive knife to the bottom. Any nearby barracuda will rush at the knife instantly, stopping just in time to avoid striking what it finally recognizes as *not* a fish in trouble.

A diver trailing a white towel like a flag—or even wearing the new white Plana or Dacor fins—is likely to be followed with puppydog devotion by any barracuda who notices the waving white. In the clear waters of Bonaire and the Bahamas, we would get the attention of barracudas (and amaze other divers with our daring) by waving our hands so that

Barracudas have an unnerving habit of following divers.

the white palms flashed. The 'cuda's eye is caught by the flash and the fish approaches until it seems to think, ''Oh, it's just some fool diver,'' and changes course. That is, the 'cuda *usually* changes course. Hand waving can become a display of underwater machismo; the waver must decide when to quit waving if the 'cuda keeps on coming. I was leading a group of divers on Bonaire's northern coast once when I spotted four small (two-foot-long) barracudas near the surface. Without thinking I began to wave my hand—and all four 'cudas began swimming toward me. They seemed to be racing each other; I didn't know if they would realize in time that the prize they were competing for was only my hand, not a succulent young fish. Choosing the safe way out, I tucked my hands under my arms. Since that experience, before I wave I make sure that only one barracuda is around!

All of this fun with barracudas must seem harmless to someone not acquainted with the fish, so a short anatomy lesson is in order. The great barracuda (*Sphyraena barracuda*) is a Caribbean fish which grows to a maximum length of about six feet—which, when distorted underwater, appears to be eight feet. However, it is not only the animal's size that strikes fear into the hearts of divers. The barracuda's teeth are formidable, noticeable, and plentiful. A shark's teeth are hidden away. The barracuda displays them every time he opens his mouth to breathe. Add the expression on the 'cuda's face—a frown—to those long, sharp teeth and to the fish's ominous manner of hovering almost invisibly or of following divers along a reef, and you can understand the unnerving effect the barracuda can have.

As with sharks, though, the barracuda's potential danger to divers is rarely actually realized (I have never heard of an incident), and certainly never in clear water where the 'cuda can easily discern between a diver and a small glinting fish.

There *is* one fish in the Caribbean Sea that is extremely territorial. It seems to have no fear whatsoever and regularly attacks animals of all sizes, including divers, which venture into its territory. As a matter of fact, the fish of this family are probably responsible for more attacks on divers than sharks

The barracuda bares its teeth every time it breathes.

and barracudas and everything else combined. Luckily, the fish in question—damselfish, family Pomacentridae—averages about four inches in size. Otherwise no one would dare to enter the sea!

These pugnacious fishies stake out their territories in stands of elkhorn and staghorn coral and defend them aggressively from all intruders. Other damselfish, the long, slender trumpetfish, the harmlessly vegetarian parrotfish, and even predators like groupers and barracudas are subject to their annoying nips. When I move in close to watch the damselfish, the little fighter will invariably attack my mask and give me a good look at the downturned corners of its mouth. Its expression seems to emphasize its aggressiveness.

It is fitting that damselfish should complete the Vertebrate Rogues' Gallery, since it is interested only in defensive interactions. Two other creatures might be considered for this Rogues' Gallery: the sharp-toothed moray eel and the venomous scorpionfish. As you've probably guessed by now, they, too, have been unfairly maligned. In later chapters, I'll show their potential for gentle interactions with divers.

Pillar coral guarded by a fiercely territorial damselfish.

VI

~~~~~~

Family Portraits

My first instructor in the art of gentle interactions between divers and marine animals was a little red hind we later named Darling. I first noticed her when a diver I was watching tried to photograph her; shy, she was too far from the camera to take up more than a small portion of the composition of the picture. I wriggled my fingers to get the attention of this small stranger, hoping it would swim closer to the camera. It moved perhaps two inches nearer, not much of an accomplishment. I held my hand palm up and wriggled my fingers again—and this time the little grouper swam right into my hand! We—the fish and I—were equally shocked, and she quickly darted out of my hand. That incident was the first of many in a lasting friendship that developed between us.

Whenever I dived Vicky's Reef, the home of this red-spotted grouper, I would see her. When I forgot to mention her in my reef orientation, divers would frequently come up exclaiming in wonder, "There was this little red fish that followed us the entire dive!" One day I guided a group of divers from the American Littoral Society. The little grouper joined us, and my buddies and I knelt in a circle on the sand as she swam up to each of us in turn, allowing herself to be tickled under the chin or stroked along her body. It was a very special experience. After the dive a young woman commented, "What a darling she is!" and my fish friend acquired her name.

77

Darling had a truly individual personality. Sometimes, on the bottom, below the boat, she and I would stare into each other's eyes for minutes at a time. Frequently she'd let me stroke her, and sometimes—the most special days—she would actually rest in one of my hands as I stroked her with the other. How wonderful to be completely trusted by a creature from an alien environment.

Darling's unusual behavior seemed to confuse the larger, more aggressive groupers. When I was feeding the fish at Vicky's Reef, the big groupers would chase Darling away so they could get more food. At those times she'd hover by my side; I'd wave a piece of food up near my face for the large groupers, and slip food to Darling with my other hand while her competitors were preoccupied. I think she enjoyed fooling those big bullies as much as I did.

But even when I wasn't feeding, the big fish would often chase Darling away from me. Their reasoning seemed to be, "If she is hanging around, she *must* be getting food. We will get rid of her and then *we'll* get the food!" They would chase her away and wait, but nothing would happen, of course.

The author's dive buddy, John D'Angelo, and Darling the grouper.

They would leave; Darling would come back. They would chase her away again. Eventually Darling's perseverance would pay off. The big fish would move away, and my friend and I could concentrate on getting to know each other better.

My willingness to interact with Darling was partly due to experiences with a grouper named Falstaff, the first fish I ever touched. I was a novice dive guide and fish feeder when we met, but Falstaff was experienced and calm. Stan Waterman was filming *A 60th at f-8*, a promotional film for San Salvador, and in one scene I was to feed Falstaff. Things worked perfectly because Falstaff knew all the right moves, and during that session he actually let me stroke him along his "cheek" (behind his mouth on his gill plate) and under his chin. That was when I realized that more senses could be used underwater than simply the visual; it was the first of my personal tactile explorations of the fish of the sea.

If we touch fish at all, it's usually as they flop around after being hooked, or as we prepare them for cooking. On the day I touched Falstaff, I fully expected him to be slimy and unappealing. Was I surprised! His cheek was scaleless and silky-smooth, but not slimy. When I tickled him under his chin, he felt rubbery. Falstaff ultimately stole the show by deserting me in favor of his own image in Stan's camera port, but my foundations for fish touching were solidly laid.

People frequently ask me, "Don't you get bored, diving the same reef over and over again?" I try to explain that diving the same reef gives me the opportunity to become better acquainted with its residents. Memorable events, like meeting Darling, or Oliver Twist swimming up off the bottom to greet me, or fish taking food from my hand, would be unlikely to happen if these animals didn't accept my presence in their world. That is my challenge in diving: to become a natural, interacting part of the undersea environment. After all, if I did not know these fish individually, I could not consider them members of my family.

Friday is a scrawled filefish, Easter blue and lavender in color, with an elongated diamond-shaped body. As a rule, filefish are shy creatures: at first, Friday was no exception to this generalization. When I took special groups to dive the

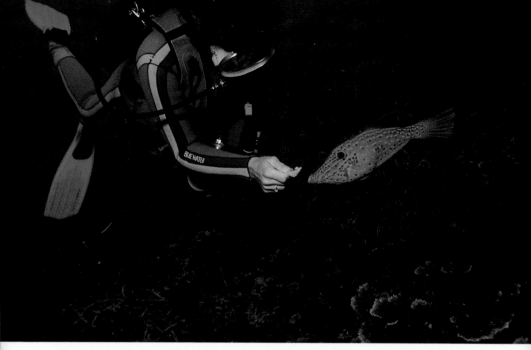

The author offering a tidbit to Friday the filefish. Photo by Tom Steinmetz.

north coast of Bonaire, I would end each dive by chumming, or scattering morsels of bait. A single scrawled filefish began swimming around downcurrent of me, nibbling at tiny pieces of fish. I waved big, succulent pieces to no avail; the filefish would not come any closer. Finally I realized that as long as the tidbits kept floating through the water, the filefish had no incentive to come closer. On my next dive there I didn't chum, but held out only one piece of fish. The filefish eyed it. I waved it enticingly. He turned his body and checked out the food with his other eye. He came a little closer, a little closer, until finally he overcame his fear and nibbled at the food. The success came on a Friday; I named the fish after the day.

Have you ever watched a filefish eat? Friday's behaviors are a wonderful reward for my perseverance in befriending him. He swims toward me as I descend to the reef—a reward

A diver can do much more than simply watch fish swim by as diver Bruce Bassett demonstrates.

The flounder is a sideways fish.   Photo by Herb Segars.

in itself, since filefish usually swim *away* from divers. Friday has funny little buck teeth with the two rows fitting together perfectly, so that when his mouth is closed he appears to have a parrotfish-like beak. He nibbles on the food I extend, then he backs off and circles around, then comes back to the food. An inveterate snacker, that's Friday. He approaches the food I offer in the same way he does his natural food of sponges and algae.

One of the interesting things about feeding fish is being able to watch *how* they eat. Eels grab the food with their teeth, hard, then writhe back into their crevices to gulp it down. Filefish and angelfish nibble. Many fish use the suction created when their mouths open to actually suck down their prey. Groupers do this, and so do peacock flounders.

I was exploring a new reef one day in May 1982, carrying my usual can of fish food, when I came upon a large peacock flounder half buried in the sand, facing upcurrent. The flounder is a sideways fish; when the fish is newly hatched, its

The author with Oliver Twist, a peacock flounder. Photo by Jeff Mondle.

right eye travels to the left side of its body and it spends the rest of its life oriented sideways, raising and lowering its pectoral fin—its equivalent of a left arm—like a sail as it swims. This flounder of the Caribbean frequently lies perfectly camouflaged in sand, but it also occasionally rests on rubble or rock. It can create lovely blue spots on its back or side or dorsal surface as the occasion demands; those spots are what earned the fish the name of *peacock* flounder. On its underside, there are no spots and there the fish is usually white.

Anyway, I was on the reef, and there was the flounder. I had tried to feed peacock flounders a few times before, with no success, but this one was positioned so perfectly it seemed a shame not to try. I situated myself, took a piece of fish from the can, and squeezed it upcurrent of the flounder. His eyes swiveled like little periscopes, and the fin around the edge of his body rippled. I waved the piece of fish and he scooted along the bottom and sucked it right out of my hand! I held out another piece of fish and he gobbled that one down, too, and then swam off.

I needed to find out if I'd met a special fish—or had just had a lucky day. I dived the reef the next day, zigzagging along, looking for that particular large peacock flounder. I found him at last, but his position was less convenient than it had been the day before. This time he was under a ledge of coral facing away from me so that I could not send the fish smell along the current to him. Nevertheless I optimistically waved a piece of food near the tail of the flounder. His periscope eyes caught the movement and he swiveled around one hundred and eighty degrees and took the food. It was crazy to believe that the flounder's behavior was deliberate, that he *knew* I had food for him, but every dive since then has supported my theory: this flounder understands!

One day I had a limited supply of fish food, and the goal of my buddy was to touch a moray eel, so we couldn't spend much time with the flounder. I fed him one piece of fish and then we continued our search for an eel. When we found one I tore a bait fish in half and offered one piece to the eel, keeping the other half in readiness in my other hand. As I concentrated on the eel, the flounder (who, unnoticed, had been following us) dashed up off the bottom and over my shoulder, stole the piece of fish right from my hand, and darted off. Remember the musical *Oliver!* and its line, ''Please, Sir, I want some more!'' I remembered it that day, and Oliver Twist the peacock flounder got his name.

What a unique animal Oliver is! His behavior constantly amazes me. One day I gave him a piece of fish tail. He took it, swished it around in his mouth, and spit out the bones in one piece! (Luckily I had a witness to this crazy observation! Oliver swallowed the entire tail, then spit out the still-connected bones just like a cat in a cartoon.) For a while I tried to feed Oliver other foods instead of fish. After some hesitation, he would suck down the hot dog or cheese—and spit it out. He has *me* well trained now: I only feed him Oliver-sized fish filets or small whole bait fish.

Obviously Oliver is more opportunistic—and thus more intelligent, in my opinion—than the run-of-the-reef peacock flounder. Right now I am interested in testing his recognition skills and his trust in me. By design, only I offer food to Oliver: I am hoping that he will at some point approach only *me*

A trumpetfish pretends to be part of a tube sponge.

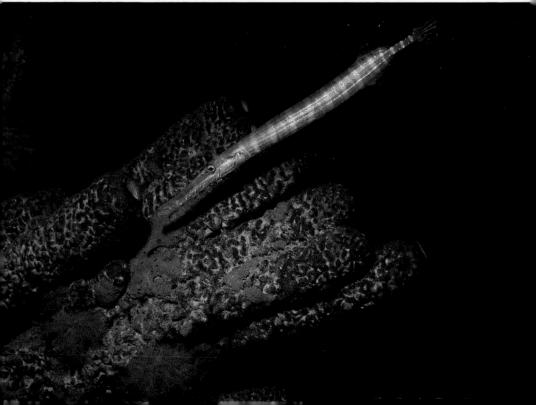

instead of my group in general. To encourage his trust, I hold food for Oliver well off the bottom, coaxing him to swim *up,* an unnatural act for a peacock flounder. Also, I always try to touch Oliver. Occasionally he will let me, but not regularly; I believe his allowing me to stroke him will be the ultimate indication of his trust in me.

Oliver Twist lives on the same reef, *Mi Cas*—"my home" in Papiamentu, the language of Bonaire—as Adelle Davis the spotted moray eel (whom you will meet in Chapter VII), and King Midas the trumpetfish. Trumpetfish are fascinating animals, and the more I learn about them the more interesting they become to me. Their long, thin bodies are perfectly adapted for camouflage; the mouths at the ends of those bodies are capable of opening to the full circumference of the bodies, so that trumpetfish can engulf prey fish that are fully as wide as they are.

I have observed trumpetfish hunting in three ways. Often they hover, head down, losing themselves among the "branches" of a feathery soft coral. Any prey that tried to hide at the base of the soft coral would get a big surprise.

Sometimes trumpetfish are seen "shadowing" other fish —groupers or parrotfish, for example—molding their bodies perfectly to the contours of the bigger (but not necessarily longer) fish as they cruise around the reef. One afternoon I was watching a trumpetfish "riding" a grouper in this manner, but when the grouper swam too close to a sea fan, the trumpetfish crashed right into the fan! He shook his head in a daze; he'd been concentrating so hard on shadowing the grouper that he never saw the sea fan. Trumpetfish shadow large fish so that the smaller fish in the vicinity cannot sense their presence. The slender predators can then dart out to gobble down the smaller fishes.

Occasionally a trumpetfish is more clever than his fellows, clever enough to exploit the activities of a diver. Such a fish was King Midas, a golden trumpetfish over two feet long. One day I was feeding fish on *Mi Cas.* Three French angelfish and fifteen or twenty wrasses were nibbling away at the hot dog I held. Then, suddenly, the wrasses were gone. They returned; they scattered again. Puzzled, I looked

A trumpetfish hiding in soft coral.

around and discovered King Midas hovering by my side. He would dart at the wrasses, they would scatter, Midas would back away, the wrasses would come in to feed again, and so forth. Rather than stalking individual fish around the reef, the clever King Midas hunted them when they were conveniently gathered together and concentrating on the food I offered! One day, in a flash of motion, he actually caught a yellowhead wrasse! The little fish struggled for its life, half in and half out of Midas's mouth—and managed to break free. My buddy and I enjoyed watching the feeding and hunting behavior, but were glad to see the wrasse escape. Midas can catch as many wrasses as he wants, as long as I am not an accessory!

Darling, Friday, Oliver Twist, Adelle Davis, Popcorn, King Midas: so far these are the only members of my undersea "family," but our interactions have taught me that the possibility for friendship with marine creatures has no limit. I dream of someday having my "own" reef, one which no one dives without me. What an incredible opportunity that would be: to explore further how relationships can develop between diver and marine animal.

A Spanish hogfish discreetly shadowed.

VII

Be Kind to Animals: Kiss a Moray

Remember the impression of the moray eel new divers got from the slide presentation? It has long sharp teeth and lurks in crevices in the coral, hoping a diver will stick his or her hand in the crevice. Then the moray eel bites down and never lets go!

The belief that eels never let go is surely the heritage from fishermen who have tried to get their hooks from the mouths of eels thrashing for their lives, suffocating in air, fighting fiercely. And the belief about eels hoping to bite divers is pure myth, at least in my experience. An eel may bite a diver for two reasons: 1) the eel feels threatened (perhaps a hand thrust unexpectedly in its face), or 2) the eel mistakes a part of the diver, such as a finger, for food. In either case, the eel will let go as long as it is not frightened—people are not exactly normal food for moray eels!

When I interact with eels, they are in their own environment; they are not threatened. A six-foot-long green moray bit me once by mistake. I held a piece of fish out to the eel, and as it headed for the fish a little grouper also went after the fish. Without thinking I waved the grouper away—and the eel, in mid-strike, shifted his aim to the movement and bit my arm. I involuntarily jerked back; the eel let go. Keeping a wary eye on the grouper, I fed the eel, who was gentle and calm, the rest of the baitfish. My souvenir of the dive was a perfect set of green moray upper jaw toothmarks in my wetsuit and my arm. The wetsuit did not heal, but my arm

A spotted moray eel.

did, so perfectly that not even a faint scar is left. The eel had bitten me by mistake, going for the motion.

Another day I had chummed a long time and was surrounded by fish-scented water. The spotted moray eel that finally responded to the chum scent was so excited that I could not get his attention onto the food—before I could move away he was wrapped around my neck and nibbling at my exhaust bubbles. He tasted one bubble right through to my neck, earning the name Vladimir, Count Dracula, for the two puncture wounds he left. Like the green moray, Vladimir nipped me by mistake.

So you see, despite their reputations, snakelike bodies, and long, sharp, plentiful teeth, moray eels are *not* eagerly waiting for any chance they can get to attack people. As a matter of fact, they can be quite gentle.

My fascination with moray eels began on San Salvador in 1979, when I was lucky enough to dive with Jacki Kilbride, who had been a dive guide on Virgin Gorda, and who was Jacqueline Bisset's double in *The Deep*. All week long Jacki kept asking me to find her a moray eel, and all week in an attempt to humor her weird request, I unsuccessfully peered into crevices and under ledges. On her last morning I found Jacki calmly stroking a three-foot spotted moray on the chin. A moray eel she had never seen before, a complete stranger! I realized you could interact with eels, and on that dive they became interesting to me.

Spotted moray eels are perfect candidates for interaction with divers. For one thing, they are plentiful. Second, each eel seems to live in a particular area. Eels aren't "territorial" in the sense that they *protect* their territories, but they remain in a general neighborhood. (I have been seeing some individuals regularly for over a year now, in the same areas.) Third, eels will allow themselves to be touched. What other characteristics could I wish for in a marine friend? I began feeding spotted morays and even a few greens as often as I could. To get the eel out of its hole a bit, I would hold the first piece of food near the eel's external nostrils, the next piece a little farther out, and so forth. Photographers diving with me had the opportunity to photograph an eel out of its hole, and I enjoyed the feeding experience.

Not a bite, just a taste. Photo by Henry and Lorisa Mansour.

One day, Elmer Munk of Elmer's Watersports, Evanston, Illinois, asked me to feed an eel so he could take photographs. We managed to find a nicely sized spotted moray early in the dive, and I began to feed it as we planned. Suddenly the eel swam away from the coral straight toward Elmer's camera. Before Elmer could even register what was going on, the eel scooted in between his camera and his BC (safety vest), then in between his BC and his body! It noticed his exhaust bubbles and ended up wrapped around his neck, biting at his bubbles from the inside out.

Being perfectly calm, I simply signaled to Elmer to stop breathing. (I could not understand why he didn't; later, when I asked him if he had understood my signal, he said, ''Sure, but how in hell do you stop breathing with three and a half feet of spotted moray wrapped around your neck?'')

With the eel wrapped around him, Elmer began, cautiously, to empty his BC pocket. Out came a spare pair of gloves, a set of decompression tables, and a flying fish he had found on the dock that morning. The eel had smelled the succulent flying fish and left my offerings in search of a juicier

The author embraced by a curious eel. Photo by Elaine Wright.

A tumor blocks Popcorn's vision.

meal, which it got, since I used the flying fish to lure it away from Elmer.

That was the first I saw of that behavior, but since that time spotted morays have wrapped themselves around my own neck several times—it is so commonplace I do not even log it anymore. (Of course, it still gets my attention!) My best explanation for the behavior is that the eel smells fish, and even with its poor vision it can see the fast-moving shiny exhaust bubbles "swimming" out of the diver's regulator. So the eel curls up in the most likely spot—around the diver's neck—and tries to catch a fast-moving shiny "fish." Eventually its lack of success discourages it and it swims back to the reef. For the diver with the eel "boa," the experience provides a sure-fire adrenaline rush. For the eel, it is probably just another day at the hunt.

Eels are often recognizable by injuries, a characteristic that puzzles me but causes me to respect these animals a great deal. We learn that "survival of the fittest" prevails in the wild, and yet eels seem to be able to overcome physical disadvantages. Recently I have been keeping an eye on Popcorn, an adult spotted moray who has what I guess is a tumor on the side of his face, between his nostril and his eye. The tumor looks a bit like a popcorn kernel, thus the name. Popcorn acts normal and looks healthy except for the growth. The tumor causes him problems since it blocks his vision on that side. Sometimes when he swims between coral heads he brushes the tumor into the coral; once it even started to bleed. Of course I have no idea what caused the growth in the first place but, eternal optimist that I am, I hope that it will heal. That might seem like a foolish hope, but another eel I met, Adelle Davis, is healing, so why not Popcorn?

An eel's natural diet consists mainly of fish, with an occasional octopus or other variation. That is why I always feed eels fish at first. Once they become used to taking food from me, most eels readily switch over to hot dogs. I try to change the food because the smell of the fish excites them and attracts other eels, and because hot dogs are easier for me to buy on a regular basis. One eel refused to switch over, and each time she spit out the hot dog I imagined her saying, "You keep your additives and preservatives, Dee, and give me some nice fresh fish." I named her after the late nutritionist, Adelle Davis.

Adelle is unusual in other ways, too. One day I noticed that her head seemed to be swollen and jellylike. I described her to a friend who told me that when people catch eels in fish traps, they bash the eels on the head—to kill them, since the eels eat the fish in the trap—and throw them back into the sea. Adelle might have been treated that way but managed to survive. As the weeks went by, though, I noticed her head swelling more and more. I began to see holes in the swollen flesh. As Adelle would take a fish from me, the holes would "smoke" with something oozing out of them. On the day Adelle uncharacteristically accepted a hot dog from me I noticed that one of her eyes was bloodshot. When I did not see her for the next week, I was convinced she had died. Then one day I passed her neighborhood and out she swam! The swelling was almost entirely gone, her skin was healed, she even refused the hot dog as usual. Adelle is my marine model of perseverance.

After many dives with her, I am beginning to appreciate Adelle's individual personality. She is the only eel I have ever known to stay in exactly the same place: under a clump of fire sponge and small coral heads. Ever since her reappearance after her illness, she has swum out of her home to meet me. On every dive she becomes more gentle and more tolerant of touching. On a dive I did with the Hassid family, she allowed herself to be caressed by Pat Hassid and her son Dan. Father Roger is convinced that Adelle's careful scrutiny of their faces was because she was looking for *me*. (That made me feel very special!) In the course of Adelle's interactions with Dan, Pat, and myself, she ended up thirty feet or so away from her home. Not wanting her to change locations because of us, I supported her and swam her back home. She allowed me to do it. Then she watched us leave, and I swear if she had had a hand she would have waved good-bye.

Interacting with animals like Adelle Davis and Popcorn is what touching the sea is all about. I have learned from them that eels can be mellow and gentle and trusting and curious. By diving with them over and over again—I have dived

The author rubs noses with a spotted moray. Photo by Jeff Mondle.

Adelle's reef one hundred and five times at the time of this writing—I am beginning to learn about eels as a group, and about Adelle and Popcorn and some others as unique individuals. The appeal eels have for me is twofold: first, it's very special to encounter an animal who will swim toward me in greeting and literally interact with me; and second, the eel's teeth and reputation give those activities an edge of excitement. I'm hoping that the day will come when Adelle or Popcorn or some of the other eels I know will clearly recognize *me* as an individual, and enjoy cuddling with me regardless of whether I bring food. When that happens I will feel justified in making a little sign for their reef that says, ''Be Kind to Animals—Kiss a Diver.''

Some words of explanation: I often refer to eels as ''he'' or ''she.'' This is strictly fantasy on my part, because the sex of moray eels cannot be determined by external characteristics of the animals. Once I name my eel buddies, it is natural to refer to them with the pronouns appropriate to the names.

Diver Alan Broder makes advances. Photo by Carole Broder.

" . . . lips I have kissed I know not how oft."

Some words of caution: Consult your dive guide before trying to feed eels! Although morays can be gentle and curious,
they are still wild animals and must be treated as such. A
diver who has seen one or two eel feedings is *not* necessarily
qualified to feed them; there are all sorts of variables that may
not show up in a couple of feedings. For example, on Bonaire
a spotted moray on *Thousand Steps* is not likely to have been
fed before, so its responses would be cautious. On the other
hand, a diver on *Sampler* or *Mi Cas* is quite likely to be surrounded by up to five free-swimming moray eels—most
sport-diving eel-feeders would be getting a little more than
they expected in that situation.

A second suggestion for those eager to feed moray eels:
wear brightly colored gloves. Fingers can look a lot like hot
dogs or even fish to eels, who have notoriously poor eye-
sight. Orange or blue gloves are obviously *not* food, and
I'm convinced they provide an additional aid to communi-
cation between diver and eel, and are thus an additional
safety factor.

A spotted scorpionfish.

VIII

The Studious Scorpionfish

One of my favorite fish is the spotted scorpionfish, partly because of its potential danger (scorpionfish have venomous spines), but mainly because it can be interacted with so easily.

What a scorpionfish wants most to do is look like a rock. To this end it rests quietly on the bottom—rocks hardly ever swim—waiting for an unsuspecting fish to venture too close, perhaps to nibble on what looks like algae growing on the "rock." When the fish gets into range, the "rock" moves very quickly; the fish disappears, and the "rock" settles back down to continue its "hunting." The reason I first touched a scorpionfish was pretty much the same reason that the small fish got eaten: I wanted to investigate the "algae," to find out if it was part of the scorpionfish or growing upon the scorpionfish. Tentatively, remembering those venomous spines, I reached out my hand and stroked the scorpionfish with one finger. I was amazed that the fish stayed in place. And I found out that those flaps were part of the fish—flaps of skin. I was also surprised to discover that the scorpionfish would allow me to touch it—until I remembered that scorpionfish are students of rocks, and rocks hardly ever swim away. The texture of the animal was very interesting. Despite the fact that I was underwater and it was underwater, it felt dry.

Continuing to stroke the fish gently, I moved my hand closer to its mouth and began to touch it under the chin. Here it felt rubbery, very much like a grouper. I maneuvered my fingers farther and farther under the fish until, to my surprise, the fish was actually resting on my hand instead of on the bottom. I gently lifted it up, one hand underneath and the other stroking it on the side. This mellow fish must have been an A student at studying rocks.

I was not actually holding the scorpionfish; I was merely providing it with a new place to settle. I continued to display it to other divers and stroke it until, uncomfortable at being so high above the reef, it swam away. It is important not to try to hold on to any animal when it tries to leave, because that is when the creature is most likely to become frightened and use its weapons.

Only once have I observed a scorpionfish in an aggressive act against a person, and that action seemed to be fully justified. In my early diving days, my buddies frequently spearfished. One day, for no reason except general perversity, my dive partner trapped a scorpionfish against the bottom with his spear. The fish was not injured; the spear was dull and the animal has bony plating protecting its head.

Some scorpionfish are fantastically ornate.

Diver Tina Marquez touches a scorpionfish, while her buddy looks on in astonishment.

Still, the fish could not have been very comfortable, and it wriggled and thrashed to get free—unsuccessfully. When Tom finally released the fish, he expected it to swim away. Instead, it swam directly toward him and did a somersault, grazing his chest with its erected dorsal spines. We both interpreted this scorpionfish's action as aggressive.

In over two thousand dives, that was the only such act I have ever observed by a scorpionfish. Most of the time the creatures conscientiously do their rock imitations until they know they have been discovered. Their actions after discovery vary. Some individuals immediately scoot away, raising their dorsal spines and displaying the brightly colored undersides of their pectoral fins as if to say, "Here I am—don't mess with me!" Others stay put but raise their dorsal spines in warning. The most mellow scorpionfish—the A students—barely move even after discovery, slowly breathing, waving their gills, as they allow themselves to be stroked.

My favorite scorpionfish earned the name Rip Van Winkle, because over a period of two months I always found

The author strokes a rocklike fellow, careful to avoid its venomous spines.

Rip in exactly the same spot. I talked with him a bit—yes, a true lunatic diver *can* croon to undersea animals through a regulator mouthpiece—and stroked him along his side. At the beginning of each of my visits, Rip was a little uncomfortable, and he raised his dorsal fin. As I spoke to him and stroked him, he calmed down and I could move a hand underneath him and pick him up. By that time he was very calm—trusting—and when I posed him for photographers he remained in whatever position I placed him in. Finally, photographers satisfied, I stroked him a bit more and replaced him exactly as I found him.

Rip lives along the shoulder of the Bonaire dropoff, about thirty-five feet deep, but I've found scorpionfish just about everywhere—usually sitting on or near the algae-covered rocks they resemble. Some scorpionfish have fewer skin flaps than others, though, and seem to be coated with fine sand and actual algae on their bodies. You might expect to find these individuals living on sand, and you will—often on the rock-and-rubble edge of a dropoff. One day I went snorkeling in a protected shallow bay with a sea grass bottom. On my way out of the water, via a tiny coarse-sand beach, I saw two scorpionfish—not lying *on* the sand but actually buried *in* it, with only their eyes and the tops of their heads showing. They are a good reason for people to shuffle their feet when wading at ''wild'' beaches.

Juvenile scorpionfish are sand-colored and usually found under rocks in rubble areas, but I have found them in one unlikely habitat: tide pools. Like the adults on the reef, these littler ones blend in perfectly with the bottoms of their tide pools. I found my first tide pool scorpionfish when my movement frightened him and he darted away (he was too young to have had much experience in rock-studying), but I found my last one when, wading along, I felt something sharp against my bare foot. I thought, ''Oops, I've stepped on an urchin or a sharp rock.'' I shifted my weight to the other foot, and looked down to see a four-inch-long scorpionfish with his first dorsal spine erect and pressed against my foot. Luckily for both of us I had not stepped down hard. Even that

The author tickles a scorpionfish. Its dorsal spines are raised in a gesture of discomfort. Photo by Armand Zigahn.

baby scorpionfish could have envenomated me—a non-fatal but uncomfortable situation—and my full weight could not have felt very good to the little guy. I moved my foot and he darted off. Apparently he felt more comfortable in his new spot (it is easy to be more comfortable without all that weight pressing down on you, I'd imagine . . .), because he let me and then my tidepooling buddy stroke him and hold him in our hands—underwater, of course. My buddy has now had the experience of cuddling with a scorpionfish—without even getting his head wet.

That's what is so endearing about these maligned creatures. Many call them ugly, and they have a prominent place in most "dangerous marine environment" lectures and books. But once I got to know them I found them to be beautiful, tolerant, and available—very special fish that have no objections to gentle interactions with people.

IX

~~~~~~~~~~~~~~~~~~~~~~~~~~~~~~~~~~~~

The Night Reef

On my first Caribbean dive trip, in 1976, my buddy and I asked the divemaster about a night dive. He pooh-poohed the idea with a challenge: "What's so great about night diving? I know how to look, so I can find anything in the daytime that you tell me you see at night." I will bet that man was afraid to dive at night, and by making us feel like inadequate daytime observers he kept his ego intact and his fears to himself. He was wrong about night diving.

Although the night reef is not a good place for interacting with reef creatures, observations there are wonderful, and some unique opportunities for tactile sensations are offered.

The reef at night is a different world from the daytime reef. Everything is the same . . . yet nothing is the same. For one thing, the stony corals are fuzzy—their polyps, usually retracted in the daytime, are extended at night for feeding. Brain corals, flower corals, star corals take on a new look. Once, when studying a brain coral closely, I found a tiny yellow goby nestled between the extended polyps, protected by them, seemingly asleep. Other sleeping fish are more easily seen: little groupers leaning against sponges, trunkfish hovering in the shelter of staghorn coral branches, blue tangs and butterflyfish in their mottled nighttime colorations resting above the sand. Within crevices in coral heads rest Spanish hogfish and creolefish and others. Under the coral heads the parrotfish sleep, protected by mucous cocoons they secreted as they settled down—it is thought that the cocoons help hide their scent from predators.

At night, flower coral polyps extend for feeding.

A butterflyfish displays its nighttime coloring.

In contrast to sleeping diurnal fish, nocturnal fish are active. Bigeye snappers hover well above their daytime lairs, feeding on plankton. Spotted drums and jackknife fish also leave their coral crevices and ledges in the safety of the darkness. Moray eels swim over and under and around coral heads, searching for unwary prey.

In Bonaire, the yellowtail snappers that cruise the reef in the daytime are still out cruising the reef at night. I have never seen a sleeping yellowtail, and I have never seen a sleeping jack, either. In San Salvador the black jacks hunt in the blue water near the reef in the daytime *and* at night. In my second year there I noticed that they had learned something from divers: they hung around us at night, waiting for our lights to illuminate some succulent animal like a little shrimp; then the jacks swooped down and gobbled up the morsels. Opportunistically, the black jacks had incorporated the activities of divers into their feeding patterns.

The fish are fun to watch, but the real reward of night diving is spotting and watching the invertebrates. During the day these animals rely on small size or deceptive camouflage or safe hideaways to elude their predators. At night, unseen in the dark, they can emerge from their hiding places. Tiny pairs of reflected lights glow everywhere, revealing the eyes of shrimps. The spiny lobster *(Panulirus argus)*, without claws but protected by short, sharp spines, leaves its coral cave at night to forage —as does the less-common shovelnosed (slipper) lobster. Shovelnosed lobsters do not have any weapons at all, certainly not claws or even spines or long antennae. In the daytime they hold on to the underside of a hollowed-out coral head and, since their colors are muted yellows and oranges and browns, even if they are seen they are not usually noticed. At night they emerge to feed and are available to be stroked and studied. Their main antennae are shortened and rounded to flat plates under their eyes, but what is most appealing about their faces is the tiny bright purple antennae located between their eyes. Both spiny and shovelnosed lobsters normally walk across the bottom quite sedately, unless they are frightened; then, with powerful snaps of their muscular (and, to hear some coarse people tell it, tasty) tails they fly backwards to safety.

The spiny lobster forages in the dark.

The nighttime reef displays its mollusks, too. The most exciting members of this group are the octopi and squid. An octopus in the daytime is likely to be in its den; at night, it is probably out hunting. Sometimes the diver's light seems irresistible to these animals. On two occasions I have had octopi swim deliberately, directly, to my light at night. Once the octopus is off the bottom and swimming, I block its path with my hands. Sometimes, as soon as it makes contact with me the octopus darts away—how strange a warm-blooded animal must feel in a cool sea! But sometimes the octopus seems to understand that we mean it no harm, and it allows us to stroke it and interact with it before it swims off.

Squid will occasionally approach a light at night, too. On night dives when my light is the brightest one around, I have attracted reef squid. One squid will swim up to within inches of the light, apparently mesmerized, but never so involved that it will allow itself to be touched. On one occasion I inadvertently approached too closely; the squid inked and fled in the same second, leaving a cloud of reddish "smoke" in the water. Octopi use their ink to create confusion: the octopus turns dark, inks a dark, octopus-shaped cloud, and darts away; the predator is suddenly presented with two octopi (one is the ink) and doesn't know which to chase. The squid, however, creates no ghost image of ink; its ink disperses immediately. Perhaps breathing the cloud makes predators uncomfortable. Or perhaps the ink provides a ''smokescreen'' behind which the squid can escape.

I was able to observe a group of many squid during some time I spent on the Bahamas bank off Bimini. At night we shined powerful lights into the water from our boat, and one night I went snorkeling to see what the light had attracted. Directly under the stern of the boat was a school of minnows that had been attracted either by our lights or by the relative darkness underneath the boat. Hovering in the light behind the minnows was a group of long, thin squid. The minnows and the squid seemed unaware of each other, but as I watched them I realized that every so often a squid would zip forward and pluck a minnow from the water—as easily as we might snatch a tomato from a grocer's display—and then

A shovelnosed lobster on the prowl.

move back into place and devour its prize. When I approached the squid, they moved away easily, maintaining a distance between us, and as soon as I backed off enough they returned to their feeding. We put pieces of raw fish into the water for the squid, but they weren't interested; they had their own natural food, and plenty of it, in that school of minnows.

Many of the lesser-known relatives of octopi and squid also come out of hiding after dark. I've found various species of nudibranchs—mollusks without shells—at night, wandering over fire coral, in pairs on algae-covered rocks, even crawling upon a sleeping scorpionfish! Flame helmet shells, which spend their days completely buried under sand, emerge after dark and move along the sand's surface. They lose their timidity at night: I placed my hand in front of one once and with only a moment's hesitation it continued its journey right over my fingers. Having a four-inch snail crawl over your fingers is a difficult-to-describe but not-to-be-missed sensation.

In the daylight, this basketstar curls up in the branches of a soft coral.

At night, the basketstar expands and filter feeds.

So if that first divemaster were to challenge me now to tell him why he should dive at night, I would have a three-part defense of night diving for him. First, although the same animals are on the reefs both day and night, their behavior is different. It is not just seeing the animals but observing their actions that makes diving worthwhile; otherwise we could all stay warm and dry and look at photographs taken by remote control. Some of the behavioral changes from day to night are radical. The orange-ball anemone is a brownish lump during the daylight hours. It awakens with the dark, expanding a ring of clear tentacles with a velvety orange ball on the end of each one. In the daytime, a tube anemone is nothing more than a hole in the sand; at night, it emerges from its tube and two circles of tentacles filter-feed nutrients from the water.

My second defense of night diving deals with the images, the visual impact of the reef at night. Because the background is black and the light I bring down shines unfiltered white light, colors are incredibly vivid. In the daytime, we see the

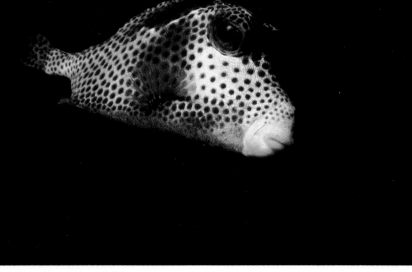

Portrait of a trunkfish.

colors of a sponge or an orange cup coral or anything else underwater with light that is filtered by the water, so the colors are tinged with blue. Using a light on a daytime dive is about as effective as using headlights when driving a car at twilight. But at night there is no background light, no light filtered by distances of water, and the full brilliance of the colors of the reef is apparent. Bonaire's *Old Pier,* for example, is an interesting daytime dive, and in the light I can scan relatively large areas while looking for particular animal friends. But the visual impact of this dive is spectacular at night, when a unique carnival of colors is created by purple and yellow or yellow and orange tube sponges and orange cup corals and multicolored filter-feeding worms and yellow zooanthids in green sponges and black-and-white spotted drums and orange seahorses and purplemouth morays.

My third defense of night diving is the most basic. I am interested in understanding the creatures of the reefs. I cannot

Orange cup corals expanded for nighttime feeding.

A sharpnosed pufferfish sleeps among the orange cup corals at night.

eliminate such a significant portion of their lives from my observations.

I hope my former divemaster reads this and realizes what he has missed all these years!

X

Touch the Sea

People frequently ask me what my goal is in interacting with marine creatures. The answer is that I want to be accepted by the creatures of the sea as one of them. Usually a diver is an intruder into the sea, zooming or crashing or hovering around, separated from the animals by something as physical as a camera or as abstract as an attitude. A diver almost always feels somewhat out of place, and too often feels threatened.

To help her scuba students overcome their fear of the sea, my friend Norine Rouse tells them, "People are in the ocean by virtue of technology, not biology. Thus we are neither the natural enemy nor the natural food source of any creature in the sea." When I first heard that, I thought, "Terrific! It really puts a diver's relationship with the sea into perspective!"—until suddenly I realized that I did not like the perspective. Oh, sure, Norine explains why we are not attacked by sharks the instant we enter the sea. What I do not like about her premise is that it emphasizes that divers are aliens in the sea who ultimately run out of air and must return to their natural environment. It is easy to understand why I was particularly excited to learn from Norine that there is at least one "legitimate" undersea interaction that a diver can be a part of: the naturally occurring cleaner station.

I first learned about cleaner stations from television shows, and then from seeing them while I was diving. In the Caribbean we have both fish and shrimp cleaners, but their functions are similar: a fish poses near the "station," and the cleaners busily travel all over their "client," removing para-

A scarlet lady, also called the redbacked cleaner shrimp.

sites from outside the body of the fish and even from inside
its mouth and gills. This process is fascinating to watch: gobies or juvenile Spanish hogfish or French angelfish or a variety of white antennaed shrimp carefully examining their
"clients," who look as if they enjoy the experience. Yellowtail snappers change in color from silver to mottled pink; parrotfish hover head up in apparent ecstasy; groupers flare out
their gill plates and turn pale, exposing their parasites for the
cleaners. Once a friend and I observed a large trumpetfish
being cleaned. The trumpetfish must have wanted something removed from its mouth, because it came back to the
station again and again, each time opening its mouth wide—
so wide that we could see the inch-long black and yellow gobies *through* the stretched skin of the trumpetfish's jaw! Animals being cleaned are perfect subjects for observation
because they stay in one place and seem less shy than usual.

As I hinted before, though, divers can do more than simply observe at a cleaner station. If we posture correctly, we

Butterflyfish, eager to be fed by Linda Kowalsky, in Molokini Crater, Hawaii.

A grouper being cleaned. The larger cleaner fish is a juvenile Spanish hogfish, the smaller is a neon goby.

The author is manicured by ghost cleaner shrimp.

can be cleaned, too. Being accepted in this way by a sea creature *as* a sea creature is a wonderful experience! I have had
my fingernails thoroughly cleaned by ghost cleaner shrimp
(*Periclimenes pedersoni*); one time a pair of scarlet ladies
(*Lysmata grabhami*) became so involved in removing dead skin
from a cut on my hand that they reopened the cut. Other
times the shrimp meticulously picked away at the hairs on
my fingers, perhaps wondering why this funny-shaped client had no scales.

For a diver, getting cleaned is not as easy as it looks. The
Fish are more finicky cleaners than shrimp; at least I have
had much more success in being cleaned by shrimp than by
fish. The only fish I have gotten to clean my hand regularly
are the neon gobies. Once, four gobies were scurrying over
my hand at the same time; on another dive, only one goby
cleaned my hand, but it got so involved that it remained hard
at work as I brought my hand within inches of my face mask
—and my noisy exhalation port. Nothing bothered that little
critter as it conscientiously examined my hand for parasites.

For a diver, getting cleaned is not as easy as it looks. The
fish clients instinctively know the correct postures that request cleaning, but we must reason them out and then experiment to see if we are correct. I have only had my hand
cleaned when it was extended thumb up; palm up or palm
down are never successful postures, perhaps because they
mimic a fish on its side, not a natural position except for
flounders. (I have never seen a flounder being cleaned, now
that I think of it; maybe their ability to bury in the sand discourages parasites.) Once my buddy extended her hand to a
cleaner station for a long time with no luck. The gobies fid-
geted and made several false starts, but they would not swim
all the way out to her hand. As I watched the scene, I noticed
that there was such a strong surge that just about everything—soft corals, hovering fish, algae—was swaying back
and forth, except the diver asking to be cleaned, who was
firmly anchored to a rock and presenting a completely steady
hand. When she gave up and moved away I took her place,
but I allowed my body to move with the surge. The gobies
seemed relieved to receive a comprehensible signal at last
and swam over immediately to clean my hand.

A goby cleans this diver's hand.

On other occasions I have had an ignored hand become cleaned after removing rings or even my watch. I've never had cleaner animals express any interest at all in a gloved hand.

I am trying to figure out the minimum qualifications that make me eligible for cleaning. Sometimes the position of my hand is enough, sometimes the fish or shrimp require my hand to be naked. I have never succeeded in being cleaned by a juvenile Spanish hogfish or juvenile French angelfish— what characteristics do I or my hand need to qualify? Must my hand have eyes? I spent one week with a fishy-looking eyeball drawn on each side of my middle finger, to no avail. Must I have scales? Gills?

I was moderately consoled by those failures to be cleaned when I learned that juvenile Spanish hogfish and French angelfish are called "facultative" cleaners; less than half of their caloric intake comes from cleaning. Since they are part-time cleaners without full instinctive "insurance," sometimes their clients eat *them!* That risk is probably what makes them avoid the unusual "hand-fish" while the neon gobies, full-time "obligate" cleaners, run less of a risk and thus can afford to be less cautious.

So far, the cleaner station is the culmination of the *Touch the Sea* experience—because it is the sea touching us, in a totally natural interaction. As more people don scuba gear to enter the world of the reef, there will always be those of us who try to understand and belong rather than simply explore and conquer. The process of learning to touch the sea is a continuous one.

Diver Jeanne Schacklett and a Nassau grouper are beginning to be friends.

Additional Reading

I hope that readers of *Touch the Sea* will be interested in learning more about marine creatures. Of the books that deal with sea stories and marine biology for the nonscientist, here are some of my favorites. Of course, there are others, but these are the ones I most frequently use and lend.

Any tales told by Eugenie Clark or Hans Hass about their early days in diving are fascinating and well written.

Reference, general:

Eugene H. Kaplan. *A Field Guide to Coral Reefs of the Caribbean and Florida* (the Peterson Field Guide series), Houghton Mifflin Company, 1982. Interesting, informative, and easy-to-read text; drawings and photographs are okay; I wish there were more of them.

Reference, invertebrates:

Patrick L. Colin. *Caribbean Reef Invertebrates and Plants*, T. F. H. Publications, Inc. Ltd., 1978. My favorite invertebrate reference book. A lot of decently reproduced, generally well-captioned color photographs. This is the most comprehensive invertebrate book I have found, and the obscure facts and in-depth descriptions more than make up for an occasional photo caption describing an "Un-known . . . ."

Gilbert L. Voss. *Seashore Life of Florida and the Caribbean*, E. A. Seemann Publishing, Inc., 1976. More scholarly than Colin with fewer photographs, but the black and white drawings are extensive and good. Voss fills the occasional gaps I find in Colin.

Warren Zeiller. *Tropical Marine Invertebrates of Southern Florida and the Bahamas Islands*, John Wiley and Sons, 1974. A beginner's book in some ways: relatively few species are shown and the text is not always very informative. Still, the animals covered are displayed in clear, well-reproduced aquarium photographs, and most of the common invertebrates are included. A good first invertebrate book.

Reference, vertebrates:

Gar Goodson. *The Many-Splendored Fishes of the Atlantic Coast Including the Fishes of the Gulf of Mexico, Florida, Bermuda, and the Bahamas and the Caribbean*, Marquest Colorguide Books, 1976. The title is longer than the book and the book is out of print, but if you can find a copy it will be well worth the search. Goodson is fairly comprehensive with good drawings and informative descriptions.

F. Joseph Stokes. *Handguide to the Coral Reef Fishes of the Caribbean*, Lippincott and Crowell, 1980. This book has interesting notes on behavior, but its descriptions are what really stand out: Stokes is designed to help fishwatchers identify "mystery" fish. The drawings are sometimes awkward (those of the spotted and purplemouth morays are my least favorites), but they are usually adequate for identification.

Ronald E. Thresher, *Reef Fish*, The Palmetto Publishing Company, St. Petersburg, FL, 1980. Although *Reef Fish* is limited to Caribbean fish of interest to the aquarist, Thresher provides fascinating information about behaviors, including feeding and mating behaviors. There are numerous illustrations and small color photographs. And who else will tell you that hamlets are simultaneous hermaphrodites?

Publications/Associations:

Most divers are familiar with *Skin Diver Magazine* and other diving-oriented publications, but too often have not heard of the following:

The International Oceanographic Foundation
3979 Rickenbacker Causeway
Virginia Key
Miami, FL 33149

The IOF publishes *Sea Frontiers* and *Sea Secrets* on alternate months. Articles cover the entire ocean-related spectrum, from marine biology to ship design. IOF has created Planet Ocean, an interesting ocean-science museum. (The bookstore there is one of my favorites.) Also, IOF sponsors trips to sites such as the Galapagos Islands.

American Littoral Society
Sandy Hook
Highlands, NJ 07732

People interested in the sea, especially if they live around the northeast coast of the United States, should consider joining the American Littoral Society, established ''for the study and conservation of aquatic life.''
The ALS sponsors local beachwalks and whale watches as well as activities in more ''exotic'' places such as the Sea of Cortez and Bonaire. *Underwater Naturalist,* the publication of the ALS, is interesting and fun.

Once you begin reading these publications and books, you will probably find yourself hanging around public aquariums, talking to (other) divers, and generally adjusting your life to reflect your interest in the sea. Enjoy!